Plant
Systematics
and
Evolution   Supplementum 5

M. Hesse · F. Ehrendorfer (eds.)

Morphology, Development, and Systematic Relevance of Pollen and Spores

Springer-Verlag Wien New York

Prof. Dr M. Hesse
Prof. Dr F. Ehrendorfer
Institut für Botanik der Universität Wien,
Wien, Austria

With 122 Figures

ISSN 0172-6668
ISBN-13: 978-3-7091-9081-4   e-ISBN-13: 978-3-7091-9079-1
DOI: 10.1007/978-3-7091-9079-1

Preface

Palynology is one of the most expanding scientific branches, both in the pure and applied fields of the natural sciences. While it certainly can stand for itself, it is especially useful as an auxilliary science in such distant fields as, e.g., botany, geology, climatology, archeology, and allergenic or forensic medicine. Still, palynological studies are far from complete despite a century of light microscope and a quarter of a century of electron microscope investigations, as stated by S. BLACKMORE in this volume.

During the XIV International Botanical Congress in Berlin on August 27 and 29, 1987, a double symposium (and, additionally, a poster symposium) was held, organized by I. K. FERGUSON (Kew) and M. HESSE (Vienna): "Spores of Pteridophytes and Pollen Grains: Development, Function, Comparative Morphology, and Evolution."

These symposia were proposed and organised because of the huge amount of new palynological information that has come to light in the last years. Specialists in several fields were invited to present their views on current research topics. Altogether, more than 40 participants offered contributions, and according to their themes the scientific sessions were grouped under three headings.

Part I: Systematic aspects
TRYON, A. F.: Evolutionary levels and ecological impact on fern spores
TAYLOR, W. C., & LUEBKE, N. T.: Spore morphology and evolution in *Isoetaceae*
NOWICKE, J. W., & MILLER, J. S.: Pollen morphology of *Cordioideae*: *Auxemma, Cordia*, and *Patagonula*
NILSSON, S.: Evolutionary and taxonomic significance of the pollen morphology of *Apocynaceae*
HUANG, T.-C.: A palynological study (TEM) of the Formosan *Apocynaceae*
GUINET, P.: The genus *Acacia* (*Leguminosae: Mimosoideae*): its affinities as borne out by its pollen characters
FERGUSON, I. K., & SKVARLA, J. J.: The pollen morphology of the tribe *Swartzieae* (*Leguminosae*)
PUNT, W.: Pollen morphology of the tribe *Dorstenieae* (*Moraceae*)
KRESS, W. J.: Form, function and phylogeny of the pollen of the *Zingiberales*

Part II: Developmental aspects
BLACKMORE, S.: Pollen wall morphogenesis and substructure
ROWLEY, J. R.: The fundamental structure of the pollen exine
OWENS, S. J., SHELDON, J., & DICKINSON, H.: The microtubular cytoskeleton during pollen development
TAKAHASHI, M.: Pollen development in *Trillium kamtschaticum* Pall.
WOLTER, M., & SCHILL, R.: The ontogeny of orchid pollinia
PACINI, E.: Harmomegathic characters of *Pteridophyta* spores and *Spermatophyta* pollen

Nine selected major papers from this symposium are included in the present volume. The topics range from skilfull investigations on the cytoskeleton in microspores (S. OWENS & al.), fascinating insights in the sporoderm substructures and unsolved questions on wall layer homologies (S. BLACKMORE; J. ROWLEY), long underrated mechanical/ecological/functional aspects (E. PACINI: harmomegathy), experimental evidence for physical features of pollen dispersal (M. BOLICK: solid and fluid mechanics in pollen transport) to questions of taxonomic (P. GUINET; J. NOWICKE & J. S. MILLER), evolutionary (S. NILSSON), or ecological (A. TRYON) significance of important pollen and spore characters. All this covers a good deal of current actuopalynology.

We are grateful to our contributors, to the co-organizer Dr. I. K. FERGUSON, to the organizing staff of the IBC programme committee, and particularly to the publisher for their effective co-operation. May this volume help to stimulate and develop palynologic research in the future.

Vienna, February 1990   M. HESSE and F. EHRENDORFER

Contents

Pl. Syst. Evol. [Suppl. 5], 1 – 12 (1990)

Sporoderm homologies and morphogenesis in land plants, with a discussion of *Echinops sphaerocephala* (*Compositae*)

S. BLACKMORE

Received December 4, 1987

Key words: *Compositae, Cynareae, Echinops sphaerocephala.* – Ectexine, homology, ontogeny, palynology, perispore, substructure.

Abstract: The requirement for an understanding of sporoderm homologies in systematic palynology is discussed. Ontogenetic investigations are considered a primary source of information for determining homologies. The developmental pathways of microsporogenesis are discussed in an attempt to determine the pattern of morphogenetic processes present in land plant groups. Perispore is identified as a layer whose relations to other sporoderm components, particularly the ectexine of pollen grains, are poorly understood. The ontogeny and substructural organization of ectexine are discussed in relation to a model that combines developmental information with concepts derived from the oxidation of exines. The elaborately structured ectexine of *Echinops sphaerocephala* is treated as an example of this model.

Systematic palynology is a discipline that exploits the diversity of taxon-specific variations between pollen grains and spores as characters that can be used for the purposes of identification, construction of classifications or phylogenetic interpretation. Although perhaps not strictly necessary in the first of these applications, hypotheses of the homology of sporoderm components are essential for the construction of classifications intended to reflect natural relationships and for purposes of phylogenetic analysis. Some systems of terminology, particularly sexine, nexine and perine (as applied by ERDTMAN 1969), are specifically intended for descriptive morphological use and enable the comparison of palynological features without implying homology between them. Other palynological terms imply homology between components that are identified by a variety of criteria including ontogeny and chemical reactivity. However, in palynology there have been comparatively few attempts to apply such criteria in establishing broad homologies between the major groups of land plants. Consequently comparisons of pollen grains and spores are generally restricted to studies within and between members of taxa of relatively low rank.

Since the introduction of the term homology (OWEN 1843) its definition, categories and the philosophical basis of comparative biology have been discussed many times (see, for example, reviews by SATTLER 1984, STEVENS 1984, ROTH 1988).

Hypotheses of homology can be arrived at in a number of ways, but one that has been widely advocated and discussed is the consideration of ontogeny (see, for example, PATTERSON 1983, WESTON 1988, DONOGHUE & CANTINO, 1988). This paper is primarily concerned with aspects of pollen and spore ontogeny and their implications for the substructural organization of sporoderms and the determination of homologies.

Comparative palynological studies at higher taxonomic levels are hampered by the difficulty of determining the homologies of sporoderm components in major groups of land plants. Our knowledge of sporoderm ontogeny, which might potentially resolve these problems, is far from complete despite a quarter of a century of electron microscope investigations (since pioneering studies by ROWLEY 1959, 1962; HESLOP-HARRISON 1962, 1963 a, b; LARSON & LEWIS 1962). Furthermore, interpretation of the diverse and complex ontogenetic processes involved is problematic and sometimes controversial. These problems are compounded by terminological deficiencies, particularly the lack of an interpretative terminology (BLACKMORE 1983) for comparing pollen grains with spores. The terminological difficulties are themselves symptomatic of the lack of consensus concerning the homology of sporoderm components.

In view of these considerations any present attempt at defining sporoderm homologies in all land plant groups will inevitably have serious limitations. However, the establishment of broad interpretations of homology can be regarded as an experimental procedure that provides hypotheses for subsequent testing and identifies particular gaps in the current understanding of pollen and spore ontogeny. ROWLEY and his collaborators have developed structural models of the fundamental organization of exines with the objective of resolving questions uof sporoderm ontogeny and homology (ROWLEY & DAHL 1977, ROWLEY & PRIJANTO 1977, ROWLEY 1981, ROWLEY & al. 1981, ABADIE & al. 1987). Recently BLACKMORE & BARNES (1987 a) attempted to identify the principal processes of sporoderm deposition and the layers that they give rise to in those algae that possess sporopollenous walls and in the major groups of land plants. In an attempt to identify critical points during ontogeny when variations can arise BLACKMORE & al. (1987) presented a phylogenetic analysis of the characters of the male developmental programme in a selection of monocotyledons with diverse types of pollen. The main objective of this paper is to extend these discussions with particular reference to the development of the ectexine. The first part of the paper is a discussion of broad questions of homology. In the second part information from a comparative study of pollen development in the *Compositae* (BLACKMORE & BARNES 1985, 1987 b, 1988; BARNES & BLACKMORE 1986 a, b) and from investigations of exines treated with potassium permanganate (CLAUGHER 1986, BLACKMORE & CLAUGHER 1987) are combined in an ontogenetic hypothesis for ectexine morphogenesis and substructure. Pollen of *Echinops sphaerocephala* L. (*Compositae: Cynareae*) is discussed, as an example, in relation to this hypothesis. *Echinops* pollen has previously been described by STIX (1960, 1964) and SKVARLA & al. (1977) and discussed in a systematic context by DITTRICH (1977). The pollen grains of *Echinops* are tricolporate with short, scattered spines and an elaborate ectexine that comprises a microperforate tectum overlying an outer zone of slender, anastomosing columellae and an inner zone of large, distally digitate columellae.

Materials and methods

Plants of *Echinops sphaerocephala* L. were cultivated at Chelsea Physic Garden, London. Mature pollen from dehiscing anthers was suspended in distilled water and sectioned with a freezing microtome and then combined with an equal quantity of intact pollen grains before acetolysis (ERDTMAN 1960). A portion of the acetolysed pollen sample was treated for 36 h with a 1% aqueous solution of potassium permanganate and then washed thoroughly with distilled water (CLAUGHER 1986, BLACKMORE & CLAUGHER 1987). Pollen grains were air dried onto stubs and sputter coated. Developing anthers representing stages from meiosis to maturity were prepared using the freeze fracture and cytoplasmic maceration technique (BLACKMORE & BARNES 1985, BARNES & BLACKMORE 1986 b). Specimens were examined in a Hitachi S 800 field emission scanning electron microscope at an accelerating voltage of 8 kV.

Results and discussion

Sporoderm morphogenesis and homologies. As products of meiosis that are surrounded by a special wall, the sporoderm, spores and pollen grains (the microspores of seed plants, CHALONER 1970, KNOX 1984) of all land plants from hepatics to flowering plants are broadly homologous. However, as FRITSCHE (1838) observed, there is only one ubiquitous sporoderm layer found in all land plant microspores, the pecto-cellulosic layer generally termed intine or endospore (KNOX 1987, BLACKMORE & al. 1987). In all spores and in the majority of pollen grains other sporoderm layers are also present. The structural intricacy of these layers and the complexity of the ontogenetic processes that give rise to them exhibit a general increase through the hierarchy of plant groups. Consequently, a comprehensive hypothesis of sporoderm homologies must be able to explain the pattern of increasingly complex ontogenetic pathways (Fig. 1) by identifying modifications, such as additions or deletions, to the development programme of microsporogenesis. At its most basic, this programme involves the sequential formation of a primary wall and a special cell wall around mother cells followed by deposition of exine and intine around daughter microspores.

The primary wall of mother cells is pecto-cellulosic but the special cell wall that unites microspores in tetrads is generally composed of callose in higher plants (BEER 1911; HESLOP-HARRISON 1963 a, b, 1968; BUCHEN & SIEVERS 1981; KNOX 1984). Callose, a β-1 − 3-linked glucan, has been reported as occurring in an almost pure form in the special cell walls of some flowering plants (HESLOP-HARRISON 1966) but is apparently absent in at least some pteridophytes (PETTITT 1971). In certain relatively derived flowering plants, including the *Compositae*, the special cell wall is highly developed and has a complex organization that has been related to its role in ectexine pattern formation (BARNES & BLACKMORE 1986 a, b; BLACKMORE & BARNES 1987 b, 1988). The special cell wall and the primary wall of mother cells are temporary structures that are dispersed enzymatically during development and are absent at maturity.

Among those algae in which sporopollenous walls occur its deposition appears to be invariably associated with distinctive structures termed white line centred lamellae (ATKINSON & al. 1972, BLACKMORE & BARNES 1987a). At its simplest, exine desposition in land plant microspores also takes place on white line centred lamellae (whose characteristics have been discussed by AFZELIUS & al. 1954, ROWLEY & SOUTHWORTH 1967, DICKINSON & HESLOP-HARRISON 1971). BLACKMORE

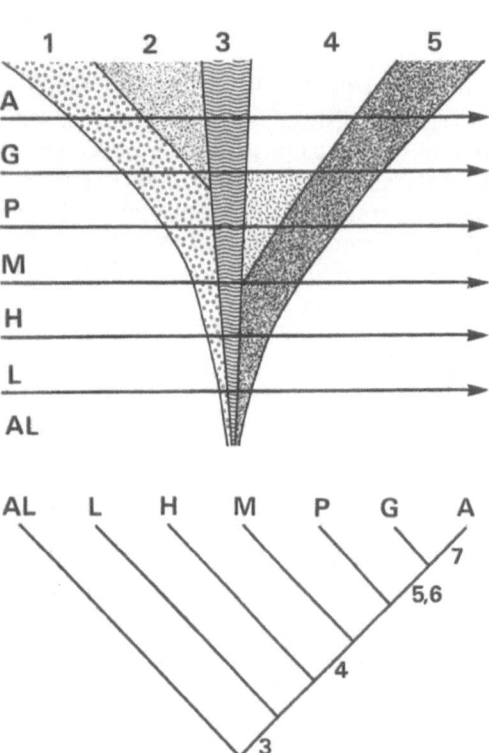

Fig. 1. Diagrammatic representation of the increasing complexity of pollen and spore development in sporopollenin possessing algae and in the major groups of land plants. For each plant group the diagram summarizes the relative time of initiation of developmental processes (from left to right). 1 Special cell wall, 2 primexine/ectexine, 3 white line centred lamellae, 4 perispore, 5 pecto-cellulosic layer, AL algae with sporopollenin, L liverworts, H hornworts, M mosses, P pteridophytes, G gymnosperms, A angiosperms

Fig. 2. The principal processes of sporoderm ontogeny shown in Fig. 1 superimposed on a summary cladogram for land plants and sporopollenin possessing algae (adapted from BLACKMORE & BARNES 1987 a). Hypothesized synapomorphies: 1 sporopollenin deposition on white line centred lamellae. 2 Intine formation. 3 Multiplication in number of white line centred lamellae. 4 Deposition of perispore from tapetal material. 5 Deposition of ectexine within a primexine. 6 Loss of perispore or its inclusion as part of the ectexine. 7 White line centred lamellae obscured at maturity. For the developmental processes of perispore formation to be incorporated in those of ectexine involves a significant change in relative timing (for explanations see Fig. 1 and text)

& BARNES (1987 a) considered that all layers deposited on white line centred lamellae originating at the microspore surface are homologous on ontogenetic grounds. Under this interpretation the inner and outer ectospore layers of bryophyte spores and at least the outer exospore in pteridophyte spores are homologous with the endexine of pollen grains. Since exospore deposition has currently been investigated in relatively few taxa (see reviews by LUGARDON 1976, 1980; BUCHEN & SIEVERS 1981) any hypothesis concerning all exospore layers is clearly open to refinement. However, sporoderm components derived from white line centred lamellae were considered plesiomorphous by BLACKMORE & BARNES (1987 a) and occur in all major groups of land plants (Figs. 1 and 2) although in many derived angiosperms they are greatly reduced or absent.

A sporoderm layer that first appears in the moss clade (Fig. 2) and is present in many pteridophytes is the perispore (the synonymous term perine is available for use in conjunction with the terms sexine and nexine). Presence of a perispore was considered by MISHLER & CHURCHILL (1984) to be a synapomorphy of the mosses and tracheophytes. Perispore is a problematic wall layer whose origin involves a considerable contribution of sporangially derived materials. The perispore is usually deposited after the completion of wall layers derived from white line centred lamellae, shortly before sporangial maturation. The ectexine is also a layer that generally involves a significant contribution of tapetally derived material but its deposition is normally initiated prior to the formation of white line centred lamellae. In comparing perispore and ectexine a substantial difference in relative timing is apparent. However, as BLACKMORE & CRANE (1968) and BLACKMORE & BARNES (1987a) have emphasized the relative timing of processes of sporoderm deposition vary considerably and do not provide an infallible criterion for establishing homologies. The mode of deposition of perispore and ectexine also differ to a greater or lesser extent. Ectexine may accumulate by several distinct processes but is deposited mainly by addition of sporopollenin precursors to a primexine (see BLACKMORE & BARNES 1987a). Primexine is deposited after completion of the special cell wall and sporopollenin precursors may be incorporated immediately, during the tetrad stage, or later, in the free microspore stage after dissolution of the special cell wall. Although the timing of ectexine deposition varies it normally takes place before the formation of layers derived from white line centred lamellae. One of the most important questions in sporoderm homology is the relationship between perispore and ectexine layers which may either be homologous or the ectexine may have an independent de novo origin which generally coincides with delection of the perispore (Fig. 1). Both layers involve a significant contribution of material from the tapetal cells and in some cases there are similarities between the mode of deposition. Recently HUANG (1987) and TIWARI & GUNNING (1987) have both applied the term perispore in descriptions of angiosperm pollen grains for layers that might more conventionally have been interpreted as the outer parts of the ectexine. Other recent ultrastructural studies have drawn attention to the presence of similar thin outer layers over the ectexine surface in a variety of flowering plants (CLAUGHER 1986, SARKER & al. 1986). Furthermore, LUGARDON (1980,

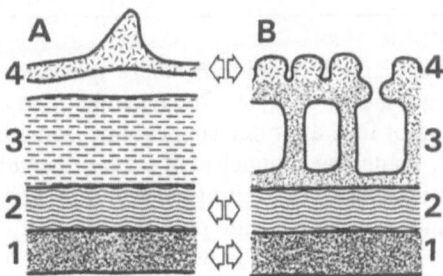

Fig. 3. Diagrammatic comparison of sporoderm layers in perispore-possessing spores (*A*) and tectate-columellate pollen grains with supratectal ornamentation derived from Ubisch bodies (*B*). Arrows and identical shading indicate components that are homologous (see text for details). 1 Pecto-cellulosic layer (endospore or intine), 2 layer derived from white line centred lamellae (inner exospore or endexine), 3 outer exospore in spores and ectexine in pollen (presumed not to be homologous), 4 perispore in spores and Ubisch body derived ornamentation in pollen

1981) concluded that the tapetal globules which build up some pteridophyte sporoderms were homologous to the orbicules or ubisch bodies produced by some angiosperm tapeta and incorporated as part of ectexine. Thus, in some cases at least it appears that part of the outer ectexine could be homologous to the perispore although a significant change in timing relative to endexine or exospore deposition would be implied (Figs. 2 and 3). Such a change would necessitate the activation of the tapetum prior to the onset of sporopollenin deposition on white line centred lamellae and the insertion of a new process, primexine deposition, into the ontogenetic programme. It is clear, however, that perispores are variable, not all are acetolysis resistant (ERDTMAN 1969, HENNIPMAN 1970) and some contain white line centred lamellae presumed to originate from tapetal cells rather than microspores. Further investigations, to determine the origin and nature of sporoderm precursors in the tapetum and the principal variations in perispore ontogeny are necessary to resolve this problem.

Ectexine substructure and morphogenesis. The treatment of exines with a variety of chemical reagents has yielded evidence for the existence of substructural features that are not usually apparent in mature acetolysed pollen walls (ROWLEY & PRIJANTO 1977; ROWLEY & DAHL 1977; ROWLEY & al. 1981; SOUTHWORTH 1974, 1986). It has always been a logical requirement that any substructures identified within pollen walls should be capable explanation in developmental terms. ROWLEY & SKVARLA (1975, 1987), for example, distinguished sporopollenin receptive and non-receptive sites within the primexine of *Canna* that define the specific form of exinous components of the mature sporoderm.

Recently BLACKMORE & CLAUGHER (1987) provided an interpretation of ectexine substructure in *Fagus* and *Scorzonera* based on the results obtained by treating mature, acetolysed pollen with potassium permanganate solutions. This treatment reduced ectexinous elements such as the tectum and columellae to a series of interconnected hollow spaces delimited by a boundary layer consisting of a fine meshwork of units about 15 nm in diameter. BLACKMORE & CLAUGHER suggested that this boundary layer might represent a feature of immature exines that can be recovered by oxidative treatment of mature pollen grains. A developmental hypothesis can explain the presence of a meshwork boundary layer within mature

---

Figs. 4 – 8. Scanning electron micrographs of mature and developing *Echinops sphaerocephala* pollen grains. – Fig. 4. Acetolysed pollen grain sectioned by freezing microtomy, showing ectexine stratification into an outer region of fine, anastomosing columellae with a distinct internal tectum below which lie thicker, branching columellae of greatest height towards the equator. × 1 000. – Fig. 5. Freeze fractured primexine during early tetrad stage, the inner face of the callose special cell wall (c) has a microreticulate patterning, longitudinally fractured primexine (p) elements show a hollow tubular construction (arrows). × 15 000. – Fig. 6. Freeze fractured primexine in the late tetrad stage showing callose wall (c) and massive primexine (p) stratified into an outer region and an inner area that corresponds to the developing large columellae. × 13 000. – Fig. 7. Acetolysed freezing microtome section of a pollen grain treated with potassium permanganate, the exinous material has been extracted reducing the columellae, tectum (t) and internal tectum (it) to a boundary layer surrounding a continuous internal space. × 8 500. – Fig. 8. Detail of an exine prepared as in Fig. 4, showing that the boundary layer is composed of material arranged in a meshwork (arrows). × 13 000

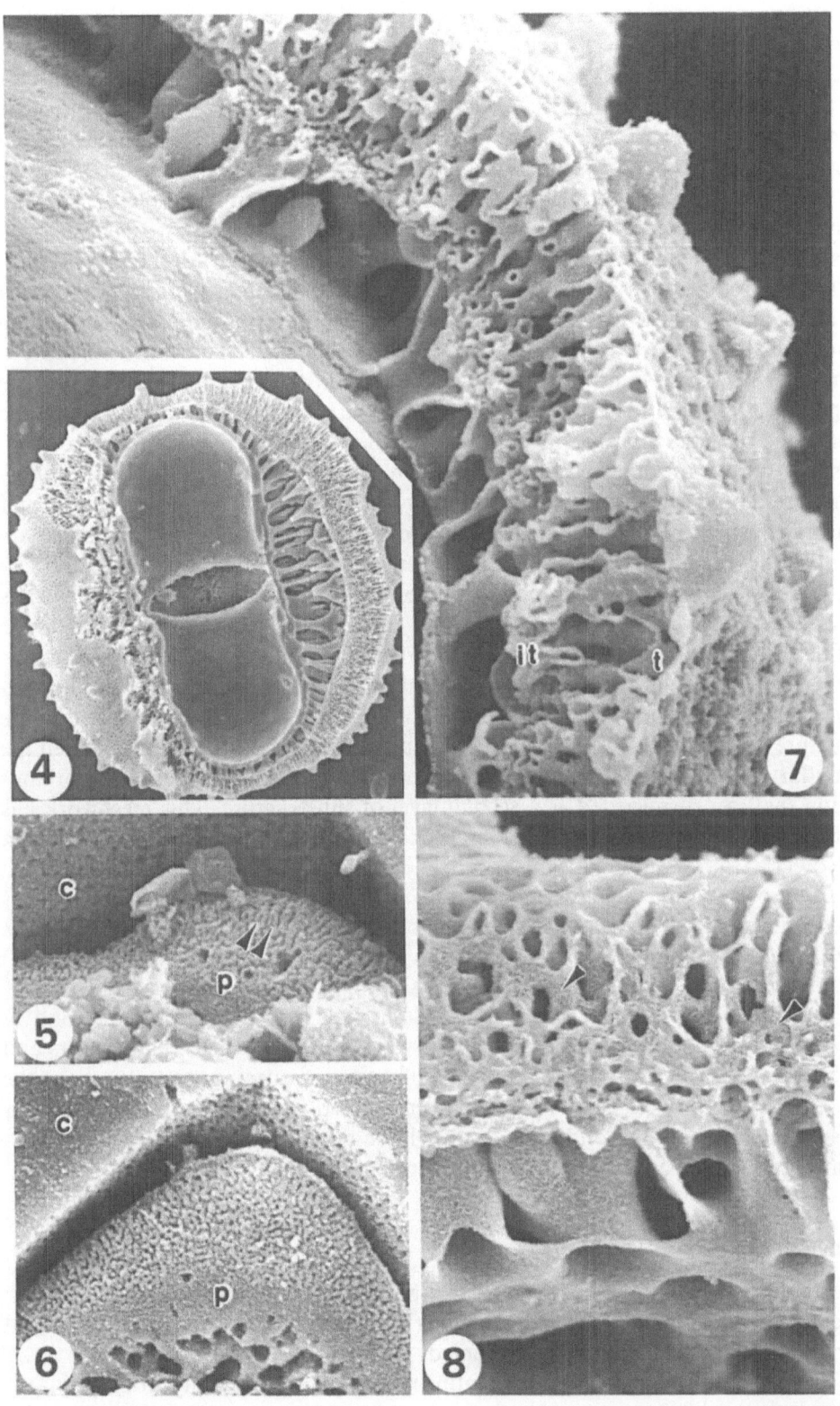

exines by interpreting it as a remnant of the primexine corresponding to the receptive sites that first define the form of ectexinous structural elements (BLACKMORE & CLAUGHER 1987). During a phase of ectexine differentiation described in members of the *Lactuceae*, for example, a skeletal ectexine is present in which the elements of the mature ectexine are present as a thin, more or less hollow columellae and tectum (BLACKMORE & BARNES 1987 b: Figs. 17–19). Such a skeletal ectexine may be widespread during early differentiation of the exine in the *Compositae,* and perhaps also in other groups. Here it is described in *Echinops sphaerocephala* which has a thick primexine, thickest in the mesocolpia, formed during the tetrad stage. In certain (radial) fracture planes the outer part of the primexine is seen to consist of a system of hollow tubes (Figs. 5 and 6) that are interpreted as corresponding to columellae in the mature wall (Fig. 4). These hollow structures, defined by a boundary layer, are interpreted as receptive sites in the primexine which become apparent as sporopollenin precursors begin to be incorporated, giving rise to the primary structural organization of the ectexine. As further sporopollenin is deposited, during the late tetrad and early free microspore periods, these hollow structures gradually become solid, resulting in the form of the mature ectexine. If mature pollen grains of *E. sphaerocephala* are treated with potassium permanganate solution the central material of ectexine elements is extracted, revealing a hollow system of spaces defined by a meshwork boundary layer (Figs. 7 and 8). Thus, as in *Scorzonera hispanica*, ectexine structure in *Echinops* can be considered to be defined by receptive sites in the primexine where a boundary layer is formed which serves to define the localization of subsequent sporopollenin deposition. The boundary layer itself can be viewed as one level of substructure that exists as a discrete structure during development and can also be exposed in mature pollen. When the boundary layer is examined in mature ectexines it possesses a meshwork organization that can be regarded as a second level of substructure. This meshwork has not yet been observed during development in the *Compositae* although CLAUGHER & ROWLEY (1987) consider meshworks to occur in many exines during development.

These observations suggest that in *E. sphaerocephala* ectexine formation involves two of the modes of sporopollenin deposition discussed by BLACKMORE & BARNES (1987 a), accumulation within a primexine (an elaborate cell surface glycocalyx) and deposition of tapetally derived material onto existing ectexine surface.

The model of ectexine substructure described here and in BLACKMORE & CLAUGHER (1987) differs from those of ROWLEY (1981) and his collaborators and of SOUTHWORTH (1986) in which systems of units are continuous throughout the ectexine. These models have, however, been derived from very different methods of specimen preparation and may eventually be reconciled through further investigation since they are not necessarily contradictory.

Conclusions

In establishing the homologies of sporoderm components it is informative to consider the principal modes of wall deposition (BLACKMORE & BARNES 1987 a). Several hypotheses may be currently be presented.

The pecto-cellulosic layer, termed intine in seed plants, is a ubiquitous wall layer in land plant microspores.

Microspore wall formation by white line centred lamellae can be considered the fundamental mode of exine deposition, producing layers that are homologous in all groups of land plants. Modifications to the programme of white line centred lamella ontogeny have included an increase in the number of units formed in land plants, the obscuring of lamellae at maturity through incorporation of additional sporopollenin (see BLACKMORE & BARNES 1987, BLACKMORE & CRANE 1988) and deletion of lamellae in derived flowering plants (BLACKMORE & al. 1987).

The relationship between perispore and ectexine although problematic is central to understanding the homologies of spore and pollen grain walls. At present there is evidence (LUGARDON 1980, 1981) for homology between the two layers in at least some plants. For ectexine to be a layer derived by the transformation of the processes of perispore formation requires significant changes in the timing of development. At present it appears probable that the origin of primexine (and consequently of primexine derived components of the ectexine) occurred within the seed plant clade. This would restrict the possible homology of perispore to those components of the ectexine that are tapetally derived without the involvement of primexine. The angiosperm Ubisch bodies which LUGARDON (1981) considered homologues of pteridophyte globules are one such component of the ectexine. Determining the origin of tapetal sporoderm precursors and their mode of incorporation in perispores, Ubisch bodies and ectexines are clearly a high priority of future research.

Ectexines of pollen grains possess substructures that relate to the way sporopollenin is incorporated within the primexine. In the *Compositae* sporopollenin receptive sites within the primexine first become apparent as a boundary layer that surrounds interconnecting hollow spaces corresponding to the tectum and columellae. The ability to recognize receptive sites at an early stage, during the deposition of primexine would enable substantial advances in our understanding of the origin of patterning in ectexines.

I thank my colleagues SUSAN H. BARNES, DONALD CLAUGHER and PETER J. STAFFORD who generously contributed to practical and theoretical aspects of this paper and JOHN R. ROWLEY for many stimulating discussions.

References

ABADIE, M., HIDEUX, M., ROWLEY, J. R., 1987: Ultrastructural cytology of the anther. II. Proposal for a model of exine considering a dynamic connection between cytoskeleton, glycolemma and sporopollenin synthesis. − Ann. Sci. Nat., ser. Botanique, 13 ser. **8**: 1−16.

AFZELIUS, B. M., ERDTMAN, G., SJÖSTRAND, F. S., 1954: On the fine structure of the outer part of the spore wall of *Lycopodium clavatum* as revealed by the electron microscope. − Svensk Bot. Tidskr. **48**: 155−161.

ATKINSON, A. W., GUNNING, B. E. S., JOHN, P. C. L., 1972: Sporopollenin in the cell wall of *Chlorella* and other algae: ultrastructure, chemistry and incorporation of ^{14}C-acetate studied in synchronous cell culture. − Planta **107**: 1−32.

BARNES, S. H., BLACKMORE, S., 1986a: Plant ultrastructure in the scanning electron microscope. − Scanning Electron Microscopy 1986, **1**: 281−289.

− − 1986b: Some functional features during pollen development. − In BLACKMORE, S., FERGUSON, I. K., (Eds.): Pollen and spores: form and function. − Linn. Soc. Symp. Scr. **12**: 71−80.

BEER, R., 1911: Studies in spore development. − Ann. Bot. **25**: 199−214.

Blackmore, S., 1983: Palynological terminology: opinion and approaches. – Grana **22**: 177–179.

– Barnes, S. H., 1985: *Cosmos* pollen ontogeny: a scanning electron microscope study. – Protoplasma **126**: 91–99.

– – 1987a: Embryophyte spore walls: origin development and homologies. – Cladistics **3**: 199–209.

– – 1987b: Pollen wall morphogenesis in *Tragopogon porrifolius* L. (*Compositae: Lactuceae*) and its taxonomic significance. – Rev. Palaeobot. Palynol. **52**: 233–246.

– – (1988) Pollen ontogeny in *Catananche caerulea* L. (*Compositae: Lactuceae*). I. Premeiotic phase to establishment of tetrads. – Ann. Bot. **62**: 605–614.

– Claugher, D., 1987: Observations on the substructural organization of the exine in *Fagus sylvatica* L. (*Fagaceae*) and *Scorzonera hispanica* L. (*Compositae: Lactuceae*). – Rev. Palaeobot. Palynol. **53**: 175–184.

– Crane, P. R., 1988: Systematic implications of pollen and spore ontogeny. – In Humphries, C. J., (Ed.): Ontogeny and systematics, pp. 83–115. – New York: Columbia University Press, and London: British Museum (Natural History).

– McConchie, C. A., Knox, R. B., 1987: Phylogenetic analysis of the male developmental program in terrestrial and aquatic monocotyledons. – Cladistics **3**: 333–347.

Buchen, B., Sievers, A., 1981: Sporogenesis and pollen grain formation. – In Kiermayer, O., (Ed.): Cytomorphogenesis in plants, pp. 349–376. – Wien, New York: Springer. [Alfert, M., et al. (Eds.): Cell biology monographs **8**].

Chaloner, W. G., 1970: The origin of miospore polarity. – Geoscience & Man **1**: 47–56.

Claugher, D., 1986: Pollen wall structure, a new interpretation. – Scanning Electron Microscopy 1986, **1**: 291–299.

– Rowley, J. R., 1987: *Betula* pollen grain substructure revealed by fast atom etching. – Pollen & Spores **29**: 5–20.

Dickinson, H. G., Heslop-Harrison, J., 1971: The mode of growth of the inner layer of the pollen-grain exine in *Lilium*. – Cytobios **4**: 233–243.

Donoghue, M. J., Cantino, P. D., 1988: Paraphyly, ancestors and the goals of taxonomy: a botanical defense of cladism. – Bot. Rev. **54**: 107–128.

Dittrich, M., 1977: *Cynareae* – systematic review. – In Heywood, V. H., Harborne, J. B., Turner, B. L., (Eds.): The biology and chemistry of the *Compositae*, pp. 999–1016. – London: Academic Press.

Erdtman, G., 1960: The acetolysis technique, a revised description. – Svensk Bot. Tidskr. **54**: 561–564.

– 1969: A handbook of palynology. – Copenhagen: Munksgaard.

Fritsche, C. J., 1838: Über den Pollen. – Mem. Sav. Etrang. Acad. Sci. Petersburg **3**: 649–672.

Hennipman, E., 1970: Electron and light microscopical observations on the perine of the spores of some *Bolbitis* species (filices). – Acta Bot. Neerl. **19**: 671–680.

Heslop-Harrison, J., 1962: Origin of exine. – Nature **195**: 1069–1071.

– 1963a: Ultrastructural aspects of pollen wall ontogeny in *Silene pendula*. – Grana Palynol. **4**: 7–24.

– 1963b: Ultrastructural aspects of differentiation in sporogenous tissues. – Symp. Soc. Exp. Biol. **25**: 277–300.

– 1966: Cytoplasmic continuities during spore formation in flowering plants. – Endeavour **25**: 65–72.

– 1968: The pollen grain wall. – Science **161**: 230–237.

Huang, T. C., 1987: A palynological study (TEM) of Formosan *Apocynaceae*. – Abstracts 14th International Botanical Congress, Berlin, p. 292.

Knox, R. B., 1984: The pollen grain. – In Johri, B. M., (Ed.): Embryology of angiosperms, pp. 197–271. – Berlin, Heidelberg: Springer.

KNOX, R. B., 1987: Pollen differentiation patterns and male function. – In URBANSKA, K., (Ed.): Differentiation patterns in higher plants, pp. 33–49. – London: Academic Press.

LARSON, D. A., LEWIS, C. W., 1962: Pollen wall development in *Parkinsonia aculeata*. – Grana Palynol. **3**: 21–28.

LUGARDON, B., 1976: Sur la structure fine de l'exospore dans les divers groupes de Ptéridophytes actuelles (microspores et isospores). – In FERGUSON, I. K., MULLER, J., (Eds.): The evolutionary significance of the exine. – Linn. Soc. Symp. Ser. **1**: 231–250.

– 1980: Comparison between pollen and pteridophyte spore walls. – Proc. 4th Internat. Palynol. Conf., Lucknow **1**: 199–206.

– 1981: Les globules des filicinées, homologues des corps d'ubisch des spermatophytes. – Pollen & Spores **23**: 93–124.

MISHLER, B. D., CHURCHILL, S. P., 1984: A cladistic approach to the phylogeny of the "bryophytes". – Brittonia **36**: 406–424.

OWEN, R., 1843: Lectures on the comparative anatomy of the invertebrate animals, delivered at the Royal College of Surgeons, in 1843. – London: Longmans, Brown, Green and Longmans.

PATTERSON, C., 1983: How does phylogeny differ from ontogeny? – In GOODWIN, B. C., HOLDER, N. J., WYLIE, C. C., (Eds.): Development and evolution, pp. 1–31. – Cambridge: Cambridge University Press.

PETTITT, J. M., 1971: Some ultrastructural aspects of sporoderm formation in pteridophytes. – In ERDTMAN, G., SORSA, P., (Eds.): Pollen and spore morphology, plant taxonomy, *Pteridophyta*, pp. 227–251. – Stockholm: Almqvist & Wiksell.

ROTH, V. L., 1988: The biological basis of homology. – In HUMPHRIES, C. J., (Ed.): Ontogeny and systematics, pp. 1–26. – New York: Columbia University Press, and London: British Museum (Natural History).

ROWLEY, J. R., 1959: The fine structure of the pollen wall in the *Commelinaceae*. – Grana Palynol. **2**: 3–30.

– 1962: Nonhomogeneous sporopollenin in microspores of *Poa annua* L. – Grana Palynol. **3**: 3–18.

– 1981: Pollen wall characters with emphasis on applicability. – Nordic J. Bot. **1**: 357–380.

– DAHL, A. O., 1977: Pollen development in *Artemisia vulgaris* with special reference to glycocalyx material. – Pollen & Spores **19**: 169–284.

– – ROWLEY, J. S., 1981: Substructure in exines of *Artemisia vulgaris*. – Rev. Palaeobot. Palynol. **35**: 1–38.

– PRIJIANTO, B., 1977: Selective destruction of the exine of pollen grains. – Geophytology **7**: 1–23.

– SKVARLA, J. J., 1975: The glycocalyx intiation of exine spinules on microspores of *Canna*. – Amer. J. Bot. **62**: 479–485.

– – 1987: Development of the pollen grain wall in *Canna*. – Nordic. J. Bot. **6**: 39–65.

– SOUTHWORTH, D., 1967: Deposition of sporopollenin in lamellae of unit membrane dimensions. – Nature **213**: 703–704.

SARKER, R. H., ELLEMAN, C., HARROD, G., DICKINSON, H. G., 1986: Recognition and response on the stigma surface of *Brassica oleracea*. – In CRESTI, M., DALLAI, R., (Eds.): Biology of reproduction and cell motility in plants and animals, pp. 53–60. – Siena: University of Siena.

SATTLER, R., 1984: Homology – a continuing challenge. – Syst. Bot. **9**: 382–394.

SKVARLA, J. J., TURNER, B. L., PATEL, V. C., TOMB, A. S., 1977: Pollen morphology in the *Compositae* and in morphologically related families. – In HEYWOOD, V. H., HARBORNE, J. B., TURNER, B. L., (Eds.): The biology and chemistry of the *Compositae*, pp. 141–217. – London: Academic Press.

12   S. BLACKMORE: Sporoderm homologies and morphogenesis

SOUTHWORTH, D., 1974: Solubility of pollen exines. − Amer. J. Bot. **61**: 36−44.

− 1986: Substructural organization of pollen exines. − In BLACKMORE, S., FERGUSON, I. K., (Eds.): Pollen and spores: form and function. − Linn. Soc. Symp. Ser. **12**: 61−69.

STEVENS, P. F., 1984: Homology and phylogeny: morphology and systematics. − Syst. Bot. **9**: 395−405.

STIX, E., 1960: Pollenmorphologische Untersuchungen an Compositen. − Grana Palynol. **2**: 41−123.

− 1964: Polarisationsmikroskopische Untersuchungen am Sporoderm von *Echinops banaticus.* − Grana Palynol. **5**: 289−297.

TIWARI, S. C., GUNNING, B. E. S., 1987: Cytoskeleton, cell surface and the development of invasive plasmodial tapetum in *Tradescantia virginiana* L. − Protoplasma **133**: 89−99.

WESTON, P. H., 1988: Indirect and direct methods in systematics. − In HUMPHRIES, C. J., (Ed.): Ontogeny and systematics, pp. 27−56. − New York: Columbia University Press, and London: British Museum (Natural History).

Address of the author: STEPHEN BLACKMORE, Department of Botany, The Natural History Museum, Cromwell Road, London, SW7 5BD, England.

Pl. Syst. Evol. [Suppl. 5], 13–29 (1990)

The fundamental structure of the pollen exine

J. R. ROWLEY

Received January 26, 1988

Key words: Pollen exine, microchannels, white-line-centered lamellations, sporoderm ultrastructure, sporopollenin, palynology.

Abstract: The "fundamental" structure, the groundwork of the exine, is a three-dimensional network recoverable from exines of pteridophyte spores and the pollen of gymnosperms and angiosperms following many different degrading methods. It can also be derived from fossil exines and untreated exines during early stages of microspore development as well. The dimensions of these networks are commonly about 70 nm measured from the centre of one lumen to the centre of the next with individual lumina being about 40 nm in diameter. My approach here is to consider the interaction between this three-dimensional network and both radially arranged microchannels and laterally arranged white-line-centred lamellations and what it can suggest to us about substructural arrangement within exines. – Through oxidation microchannels can be hollowed out to diameters of 40–70 nm indicating that three-dimensional networks are superpositioned around microchannels. – White-line-centred lamellations have two features of exceptional interest with respect to how they pass through the three-dimensional network. They are both wider than the meshes of the network and apparently come and go during development. I consider the endoaperture of *Epilobium* to be a useful model in this interpretation. Exine unit structures are attached to either side of white-line-centred lamellations in these endoapertures; using this system as a model I suggest that white-line lamellations can be junction planes between units structures. The important feature of my interpretation is that all substructures of exine units take part in the white line structure including those appearing after partial degradation of the exine as a 3-D network. In this model white-line lamellations can be transposed into subunits of rod-shaped unit structures and the reverse. Sketches of this reversible system were prepared for the THANIKAIMONI memorial volume of the Journal of Palynology.

Introductory background

Before degraded exines completely disintegrate they often appear electron microscopically as a three-dimensional network. I call this three-dimensional (3-D) network "fundamental" because it seems general in occurrence in distantly related taxa and may be common for all exines. There are many published examples of similar 3-D networks (e.g., EHRLICH & HALL 1959; many papers by AUDRAN who obtained netted exine remnants through oxidative treatment, i.e., AUDRAN 1970, 1977, 1980; KEDVES & PÁRDUTZ 1970, 1983; SOUTHWORTH 1985 a, b, 1986 a, b; see, however, qualifications represented by quoted descriptions below; for additional citations see legends for Figs. 1 – 6). The micrographs in Figs. 1 – 6 represent four major groups, *Lycopodium*, a pteridophyte and *Artemisia*, a dicot angiosperm (in Figs. 1 and 2); Fig. 3 *Classopollis*, an extinct conifer exine from Upper Jurassic; Fig. 4 *Lilium*, a monocot angiosperm; and Fig. 5 *Thunbergia* and Fig. 6 *Betula*, dicot angiosperms. I see these three-dimensional (3-D) networks as interdigitations between binder subunits around neighbouring rod shaped exine units. While I refer throughout this paper to images like the above as "3-D networks", "networks", or "nets", exine remnants do not all look exactly alike nor are they described in the same way. In describing partly degraded exines of *Lilium* SOUTHWORTH (1986 b) emphasized the granular aspect, "a lattice-like structure of interconnected granules" and "irregular polygons of interconnected granules" (see her observations of *Fagus* and *Juniperus*) (Fig. 4). In the earliest modelling of exine substructure AFZELIUS & al. (1954) considered regularly aligned granules (c. 1 nm in diameter) to be substructural components of sporopollenin. KEDVES (1986) has suggested a globular structure for the biopolymers of sporopollenin in *Corylus avellana* based upon enzymatically degraded exines.

While the substructure of intact exines is unlikely to be understood directly from these 3-D networks alone because too much is missing, the "nets" conceptually provide much that is constructive both with regard to the nature of exines and questions they force us to consider. I emphasize two questions in this paper; how microchannels and how white-line lamellations fit into these networks, but there are, of course, many others.

One obvious question concerns the extent to which these networks could be reconstitution phenomena involving reassembly or self-assembly of products of exine degradation.

A major point seeming to verify the existence of networks comes from developmental studies. Some of the published descriptions of proexines and exines of early free microspores include comments about spiderweb-like arrangements of microfibrils (ABADIE & HIDEUX 1979, HIDEUX & ABADIE 1985), honeycomb-like patterns during exine formation (EL GHAZALY & JENSEN 1985), and fibrillar networks in the exine (ROWLEY & DAHL 1977: e.g., Pl. 23, Figs. 1 and 2). Other examples are reviewed by SOUTHWORTH (1985 a), emphasizing upon granules rather than connecting filaments. This emphasis is consistant with the observations of GABARAYEVA (1986, 1987), WAHA (1987), and with the early stages reported by

---

Figs. 1 – 3. Spore and pollen exine structures. Bars: 100 nm. – Figs. 1 and 2. The upper portion of Fig. 1 is an exine remnant of *Lycopodium clavatum* L. while the lower part and all of Fig. 2 is *Artemisia vulgaris* L. These exines were treated together and shown in the same micrograph at lower magnification in an adjacent section (ROWLEY & al. 1981 b: Fig. 6). A 3-D network (circles) is shown in both exine remnants. When viewed on end lacunae of the network are pentagonal or star-shaped (circle – p). Subunits that are straight running through the network in *Artemisia* are marked by arrows. Treatment: GEA to 2-aminoethanol to potassium permanganate to plastic embedding to thin sectioning to plastic removal to negative staining. – Fig. 3. *Classopollis classoides* PFLUG exine in transverse section negatively stained following removal of embedding plastic. This micrograph of an extinct gymnosperm exine is a portion of a figure from ROWLEY & SRIVASTAVA (1986). The inserted sketch offers an interpretation of the loops in core subunits surrounded by the darkly contrasted binder subunits. The exine was recovered from a Jurassic outcrop

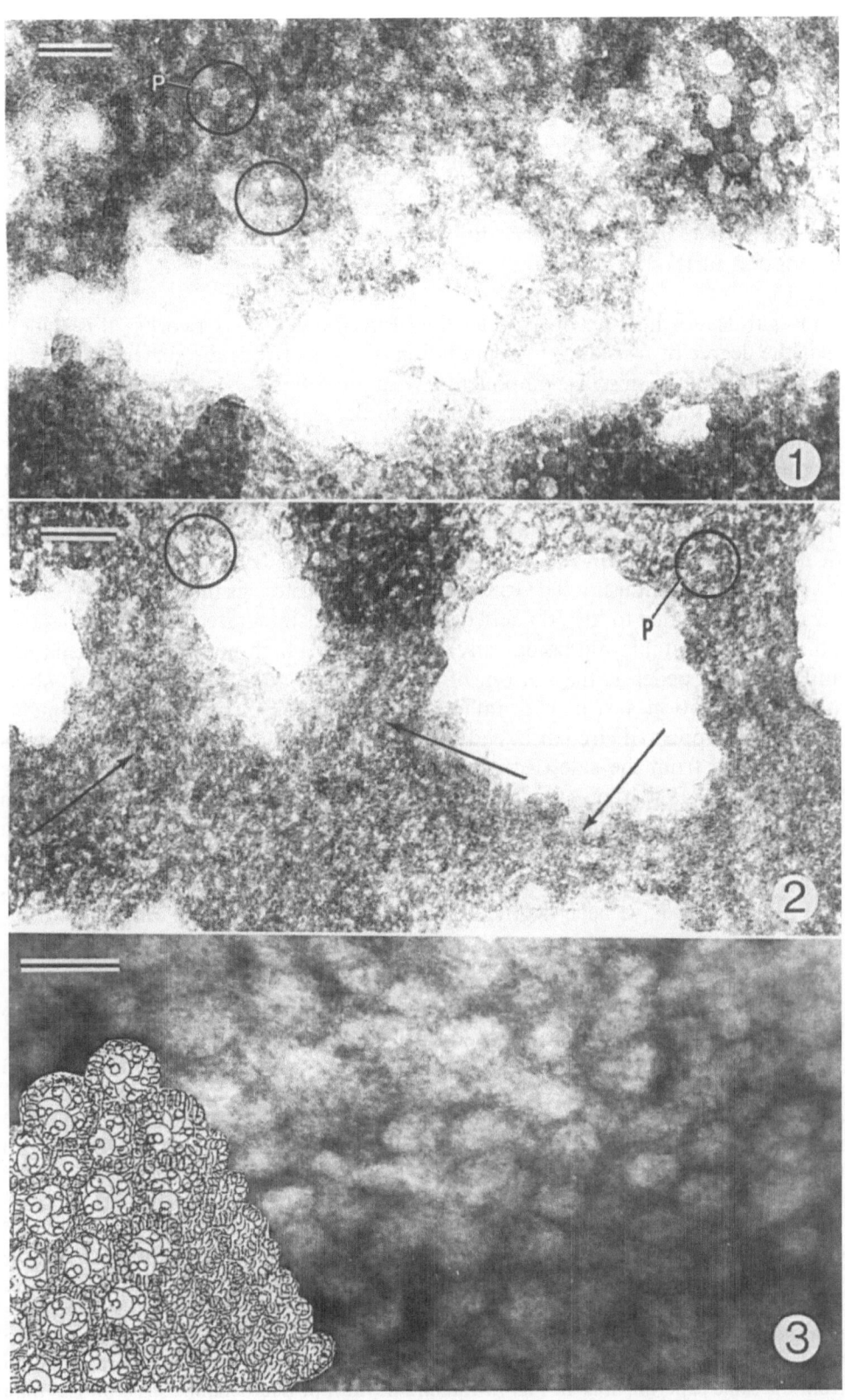

ROWLEY & SKVARLA (1987). WAHA refers to exine receptors as "ellipsoidal sporopollenin precursor particles".

My conclusion is that the granules are part of the core of exine units. The core is surrounded by a zone of binding subunits that do not accept stain during some periods of development. (This is the condition in the early exine of *Fagus* in Fig. 13.) When the binding zone is stainable there appears to be a stain reversal (for micrographs and sketches of such stain reversals see DUNBAR & ROWLEY 1984); there are examples of intermediate conditions in ROWLEY (1987). These points will be elaborated further in connection with discussion of the filamentous network in Fig. 13.

Regardless of how one interpretes the form of the "3-D networks" it is evident that the degree of resistance to degradation differs between the surviving remnant and the apparently missing components. What does this mean in terms of structure and composition? Sporopollenin is known to be degraded by oxidation and these net-like remnants are in, at least, many cases what is left of exines after that treatment. Since the net or parts of it resist GUNNAR ERDTMAN's acetolysis (1960), as in Figs. 1, 2, and 5, these residual 3-D networks are, by current definition, sporopollenin. (A precise method involving immunological techniques for identification of sporopollenin is described by SOUTHWORTH & al. 1988.)

In an experiment using the GUNNAR ERDTMAN acetolysis method (GEA) exines previously degraded to the 3-D network level were disintegrated by GEA (no intact exines were detectable microscopically); nevertheless a high-mass remnant remained and formed a pellet at the bottom of the centrifuge tube after low speed, short stop centrifugation, just as with intact exines. In TEM micrographs components in the pellet appeared circular in end views and to have a coiled or zig-zag aspects when viewed from the side (ROWLEY & al. 1981 b: Pl. 8, Fig. 27). The remnants SOUTHWORTH (1986 b) observed from the "bottom of the tube" after disintegration of pollen exines were described differently. Her micrographs include fragments in the form of "solitary granules 10 – 15 nm in diameter or associations of bent or branched chains, circles or double rings" (SOUTHWORTH 1986 b). We agree, however, that after exines have been degraded to the "3-D network level" and then disin-

---

Figs. 4 – 8. Pollen exine structures. Bars: 100 nm. – Fig. 4. Section of a murus and subtending bacule in exine of *Lilium*. The exine was treated with 2-aminoethanol (from SOUTHWORTH 1985 a). The dark bands at the base of the bacule are referred to as surface coat and intine. – Fig. 5. Scanning EM micrograph of the exine of *Thunbergia mysorensis* T. AND. after GEA to potassium permanganate (from CLAUGHER 1986). The bars of the 3-D networks show a texture which shows cross striations in some sites (arrows). Seen on end the meshes appear polygonal (circles). – Fig. 6. Tectum of *Betula verrucosa* L. exine eroded in a fast atom source. The hollow cylinders (arrow) in the cairn-like remnants are exine "unit" structures under spinules. The loops in the background are considered to be eroded cylindrical unit structures. The core zone noted by the arrow is comparable in size to the dark stained-tunnelled out microchannel in Fig. 7. Treatment: Acetone to a fast atom source for 30 min. – Fig. 7. Microchannels (arrows) in tectum of *Betula* pollen after oxidative treatment. The reduction in tectal (T) height and nexine (N) thickness as compared to the section in Fig. 8 may be due to etching by oxidation. – Fig. 8. *Betula* exine before oxidation from the same experiment as in Fig. 7. Maximum width of microchannels is a bit less than 20 nm

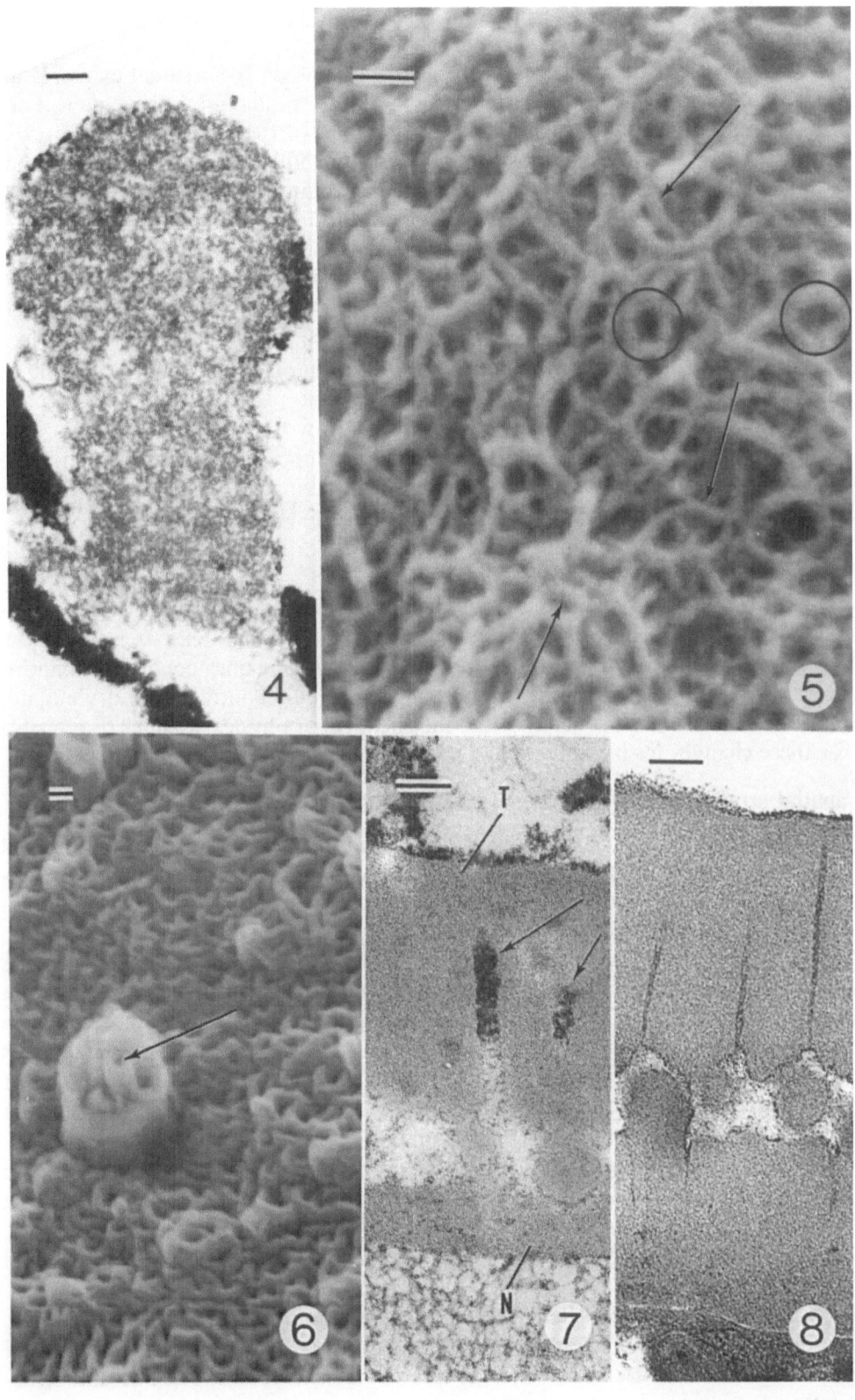

tegrate there is a remnant similar in form to components seen prior to exine disintegration.

Why if sporopollenin is present in these 3-D nets do the residual exines disintegrate in hot acid? My explanation rests upon the results of experiments BOTJAH PRIJANTO and I did with attention to observations of the late ROBERT TSCHUDY. TSCHUDY (pers. comm.) had observed numerous examples of subfossil exines that expanded enormously under conditions such as dilute hydroxide that had no detectable effect upon most exines. PRIJANTO and I found that the actual loss of more or less intact exine form did not occur in acid but rather on going to water or water followed by dilute hydroxide where the degraded exines swelled and were subject to disintegration (ROWLEY & PRIJANTO 1977). It is my conclusion that disintegration of the "fundamental" exine network results from expansion and fracturing of the structurally incomplete exine remnants rather than by etching or dissolution.

A basic question remains as to why a remnant with expected features of sporopollenin (resistance to GEA and high refractive index, cf. ROWLEY & PRIJANTO 1977) is left after oxidation, especially a remnant having such an enormous surface area as a 3-D network. SOUTHWORTH has suggested the idea of limit-sporopollenin for exine remaining after destructive treatment, patterned after limit-glucose (D. SOUTHWORTH, pers. comm.).

While the emphasis above is upon expansion and disintegration, many exines degraded to the 3-D network level can contract to a "normal" size after expansion. ROWLEY & al. (1981a) found that treated and control exines could be expanded in water or dilute hydroxid many times and repeatedly contracted in dry ethanol or dilute acid. It can be concluded that the chemical or physical systems responsible for these changes in size are retained in exines degraded to the 3-D network level.

Spatial arrangement of the 3-D networks as indicated by microchannels

One approach to finding out how missing components may have been arranged in 3-D networks is to consider how structures commonly called microchannels pass through the 3-D networks. Microchannels are relatively long structures, often radially orientated.

We set up conditions which might be expected to tunnel out microchannels (ROWLEY & al. 1987). What happened was that the treatment designed to bring about oxidation of exines enlarged microchannels, in so far as stain acceptance was concerned, to equal the normal diameter of meshes of the 3-D net. Examples of sections of oxidized and control material occur in Figs. 7 and 8. In examples of the widest oxidation-induced microchannel tunneling a coiled or zig-zag component (Fig. 7) appeared at the level of the "filaments" of the 3-D network.

In Fig. 6, *Betula* pollen from the same collection as Figs. 7 and 8 had been exposed to a fast atom source (see CLAUGHER & ROWLEY 1987, for methods and discussion). The cairn of large (c. 100 nm wide) cylindrical structures in Fig. 6 is located in the tectum etched out from under one or more spinules and a tectal ridge. Each "cylinder" is considered to be a substructural unit of the exine. The dark central core in one of these units is about 40 nm in diameter – a little less than the diameter of the tunneled out microchannel in Fig. 7. The net-like background surrounding the cairn of exine units is also thought to consist of exine unit-structures.

In conclusion, the 3-D network appears to go around microchannels with lacunae of the net aligned radially in superpositioned accordance, like a wall of stacked and nested cylinders. It also seems from these results that the outer "wall" of tunneled out microchannels is part of the 3-D network. Based upon exines etched in a fast atom source, as in Fig. 6, it looks as though a 3-D network could include the outer "binder" portion of cylindrical exine units.

How white-line-centred lamellations fit into a 3-D network

White-line-centred lamellations (white lines) offer a more challenging problem than microchannels, for when they are wider than c. 40 nm they might not fit into or pass through a 3-D netork. Here I consider white lines as they occur through the endexine of *Epilobium angustifolium* where they are up to 200 nm in width and at least several micrometers in length.

There are not many measurements of the widths of white lines. HECKMAN (1970) determined that they were c. 120 nm wide and 250 nm long on the surface of spores of the liverwort *Lophocolea*. Because they were thin and rectangular she called them slips. At the surface of the sporoderm of *Canna* we measured lamellar slips with a white line centre to be c. 70 nm wide by 500 nm long. In the endexine of *Grevillea* (*Proteaceae*) white line lamellae were c. 50 nm wide (ROWLEY 1975). GOD-WIN & al. (1967) in one of the first descriptions of white lines, in *Ipomea*, termed them tapes because of their proportions. Tapes and slips are interesting descriptive terms and suggest that white line forms may be distinct in various taxa. The ultimately simple white line system, and a kind of proof that all white line systems are not the same, has been recorded in the endexines of the legume *Poinciana* and the extinct conifer *Classopollis*. In these taxa white lines are based upon a tubule central in a rod structure (ROWLEY & SKVARLA 1987, ROWLEY & SRIVASTAVA 1986).

Based upon serial sectioning my estimates of white line widths in *E. angustifolium* range from 40 nm (often entirely within one section) to 200 nm. Because there are no reference points for white lines within the endexine the serial section method was not entirely satisfactory. The widths of white lines in my samples were greatest during periods of physiological activity, represented by irregular channeling as in Fig. 9, and the early vacuolated period of microspore expansion (see ROWLEY 1988). In sections showing white lines in relatively long profiles (0.5 – 1 μm or more) there will usually be a region of increased width and decreased whiteness, representing, presumably a "half twist" in the lamellation. The incidence of "twisting" and, also, the number of white lines per unit-area of section profile is highest during intervals of hyperactivity such as the uptake associated with the bulged endexine in Figs. 9 and 10 (see ROWLEY 1988, for results of tracer experiments). In all stages of development it is most common for white lines to be present at the ectexine-endexine junction (see white line marked "Jp" in Fig. 12 and similar location in Fig. 10).

For *E. angustifolium* there are published micrographs of the endexine appearing as a 3-D network with no obvious meshes much larger than 40 nm after a variety of methods (ROWLEY 1971: Fig. 15, ROWLEY & PRIJANTO 1977: Figs. 5 – 7). The network in the endexine of *Epilobium* is similar in appearance to the image in Fig. 5 (Fig. 5 is a small portion of a micrograph in CLAUGHER 1986). A network has been seen within the endexine of *E. angustifolium* without embedding resins using high resolution SEM (D. CLAUGHER, pers. comm.).

Assuming that the network is not reconstituted during or after exine degradative procedures, white line lamellations much wider than 40 nm can be expected to form a part of the 3-D net rather than passing through aligned meshes like a ribbon. There appears to be a solution to the way white line lamellations are arranged in the exine that can be seen in diagrammatic form in endoaperture regions of young *Epilobium* microspores (Fig. 12).

In the endoaperture regions white lines project from an apparently solid endexine in sections like Fig. 12 (see also Rowley 1988). There is a very well ordered fringe on either side of the white line-centred lamellation which I interpret to consist of short cylindrical exine unit structures. The cylindrical unit structures are nested as they attach to these white lines, and in end view the contours of neighbouring walls of the units are square or polygonal (see circled sites in Fig. 12). Such contours I consider to be portions of a 3-D network based upon nested unit structures. Parts of a unit structure marked "B" and "C" in the insert to Fig. 12 are comparable in shape with circled sites in Fig. 5 and sites marked "circle − p" in Figs. 1 and 2. The isolated unit structure in Fig. 13 marked "B" and "C" is similar in size to the above cited structures. The binder zone "B" of the unit in Fig. 13 shows several lines of contrast as is also apparent throughout the rest of the forming exine and in the bacular arcade. These lines of contrast result from stain-accepting material in and on substructure (see Rowley 1987). These lines of contrast form a fine mesh. According to my view the relatively coarse 3-D networks (e.g., Fig. 5) are the result of the sum of subunits in the binder zone whereas the fine mesh throughout Fig. 13 is produced by components associated with individual subunits. Images in Figs. 1 − 3 are likely to result from some of both.

The unit structures in the endoaperture region have an uncontrasted core zone resembling that of the processes in the ectexine of *Epilobium* (Figs. 9 and 12); see also micrographs shown by Keri (1986) representing many different stages in development. Reversals in contrast in the core and binding zones of unit structures occur in many taxa (see review in Dunbar & Rowley 1984). A condition where a dark core is surrounded by a light binder-zone is seen in Fig. 13. In the endexine of *Epilobium* the reversal may occur repeatedly as can be appreciated by the occurrence of low contrast sites 40 nm wide in bulged regions (Figs. 9 and 10).

My conclusion is that the kind of cross-wise association between exine units and white lines seen in the endoaperture region also occurs within the solid appearing endexine. There are examples in Figs. 10 and 12 where the radial processes of the ectexine join the solid endexine at a white line. Such sites are marked "Jp". In this

Figs. 9 − 11. Pollen exine structures of *Epilobium angustifolium* toward the end of the vacuolate period. The three figures could be obtained from different sectors of one microspore. − Figs. 9 and 10. Sections of progressive aspects of regions of recession of irregular channeling. There are many white lines in both figures but in Fig. 9 "twisted" profiles are up to 200 nm wide (arrow heads). In both figures there are circular (rod-shaped) profiles low in contrast (arrows) that are like the holes in processes of the ectexine (Figs. 9 and 12) and endoaperture (Fig. 12); these rod-shaped 40 nm wide profiles connect to white lines whereas irregular channels do not. − Fig. 11. Thin section negatively stained after removal of embedding plastic. There are lateral lines of contrast which could be white lines. The most prominent features are radial (e.g., between arrows) from base to top of the nexine and short lateral or oblique lines (prominent within circle) across the radial structures

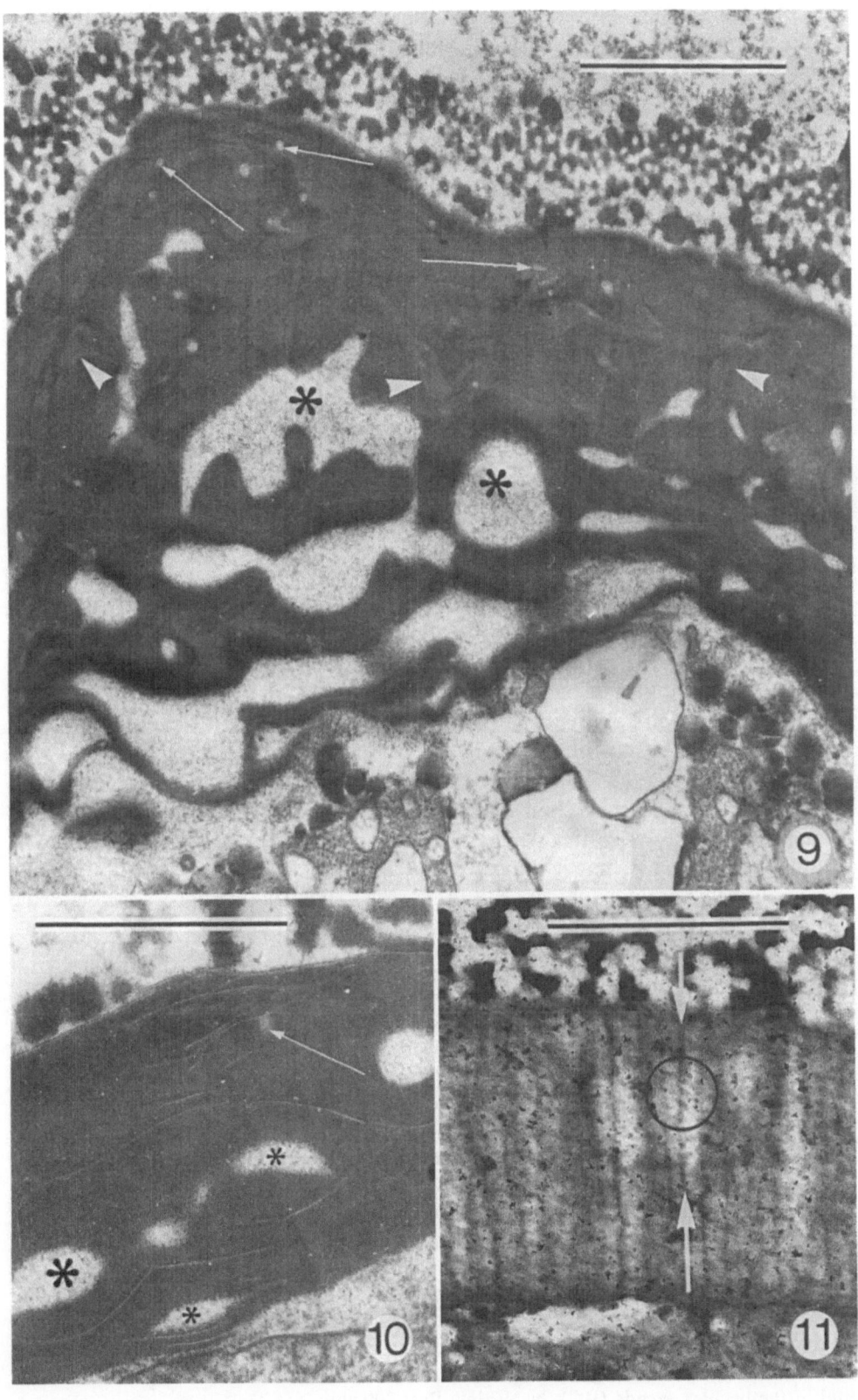

kind of white line system the white line is a junction plane within the endexine, a structural lamellation allowing for deformity and followed by structural reassembly of endexine form.

The endexine of *Epilobium* does in fact undergo enormous deformation. I have referred to the structural aspects as "irregular channeling" within bulged regions of the endexine (Rowley 1976). The following is a summary of some conditions associated with irregular channels. The bulges are enormous, commonly 3 µm in height, whereas the endexine normally is 0.7 µm thick with irregular channels more than 10 µm wide. They close up when the temperature is lowered below 4 °C, indicating an energy requirement. They also close during fixation with aldehydes and so will not be apparent in many fixed materials. Closing-up of channels as observed by light microscopy occurs within 10 min; after that, in TEM preparations there are only white lines left to indicate channeled regions, and after 30 min no white lines are in evidence. This is a very dramatic example of endexine activity and must be taken seriously in consideration of exine function. Results of the extensive studies of Kurmann (1986) suggest that "irregular channeling" in the lamellated wall zone may be a common feature during development of gymnosperm microspores. The interbedding of a fibrillar matrix within endexines noted by Skvarla & Larson (1965) may be an indicator of irregular channeling.

When the irregular channels are open as in Fig. 9 there is a white line in about the middle of each "wall" around a channel. This is similar to the endoaperture "model" in Fig. 12 where there are channel-like open spaces (asterisks) similar to the irregular channels marked with asterisks in Figs. 9, 10, and 12. (Note that this channeling does not occur on white lines.) Recovery from the bulged and irregular channeled condition apparently involves reunion of exine unit structures with no evident structural disjunction (see Fig. 11).

The bulge in Fig. 9 is in recession when it was common to see swirled lines of nonuniform stain acceptance in the endexine. These are traceable to white lines. The importance I give to this swirling of white line structures within the endexine at times of bulging is that it indicates that the most astounding rearrangements occur routinely within exines since at other times substructure across the endexine may become extraordinarily well aligned radially (Fig. 11).

Figs. 12 and 13. Exine structures of *Epilobium* and *Fagus*. – Fig. 12 and insert. Endoaperture region of *Epilobium angustifolium*. The complex processes of the ectexine (S) join the "solid" endexine (N) at a zone (Jp) that is similar to the zone to either side of white lines in the endoaperture (region of curved arrows). The circled sites emphasize the unstained rods that join white lines as they are seen in end view. The unstained rods and the "walls" between them are attached only at one end to the white lines. The spaces at the other (unattached end) of these exine unit structures are like irregular channels (marked by asterisks here, and in Figs. 9 and 10). Inserts show enlarged images from regions marked by a long line and arrows. The curved arrows indicate places where subunits coil around white line while straight arrows indicate profiles of c. 10 nm wide subunits in region where a white line is "twisted" and increased in width. C Presumed core, B binder zones of unit structures. Bars: 100 nm. – Fig. 13. Early free microspore stage for *Fagus silvatica* L. Finally the exine unit structures (C) are the reverse of unit structures in Fig. 12. The core zone here is dark (C) and binder zone (B) very lightly stain accepting. There is a network throughout the exine and in the bacular arcade (BA). N Nexine. Bar: 200 nm

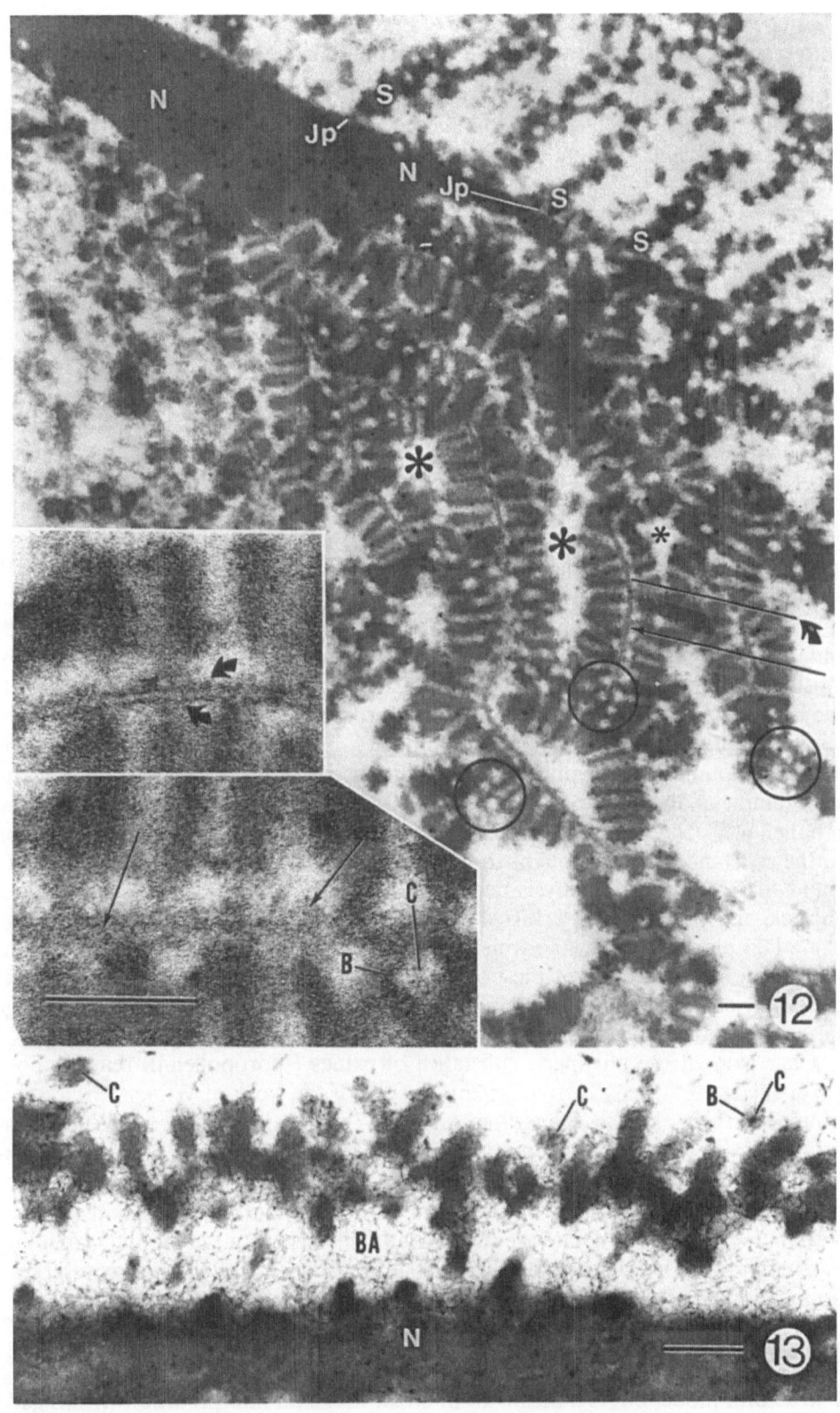

Missing components from lacunae of 3-D networks
Difference in chemistry and structure of core and binder subunits

Considerable attention is devoted above to 3-D networks as remnants of exines lacking components such as microchannels and white lines. I suggest that the missing components formerly found in lacunae of the networks become fractured and mechanically dispersed without actually being dissolved away. In other words the missing parts of the exine fracture, disintegrate, and then fall to the bottom of the tube in the same way as is proposed for disintegration of the 3-D network when exine remnants totally come apart. After the oxidation treatment resulting in the exine condition referred to here as a 3-D network low-speed short-stop centrifugation produces a heavy residue. The freeze-dried weight of the 3-D exine network plus the "bottom-of-the-tube" material was about the same as the freeze-dried weight of the exines after acetolysis.

My experience with processes on tapetal cell surfaces (connected by viscin threads to exine processes) has been that core and binder zones react differently to stains and that stain accepting substances are extracted at different rates from core and binder sites (ROWLEY 1987). These are likely to be reactions of sporopollenin precursor macromolecules (components of the glycocalyx) rather than reactions of sporopollenin. My view is that the plasma membrane glycocalyx components which form the template for the exine remain encapsulated within sporopollenin for as long as the exine lasts (ROWLEY 1975, 1978). When the exine has been eroded as under conditions resulting in 3-D networks, contraction or expansion of the encapsulated carbohydrates may mediate enormous change in size of exine remnants and/or disintegration of either all or some specific components of the exine.

The differing resistance of core and binder subunits may actually be caused by varying amounts of sporopollenin, or by impurities in the sporopollenin. Silicon, one example of an "impurity", was reported to be a prevalent elemental component in pollen wall structure by CRANG & MAY (1974).The effect of elements like silicon on the resistance of exines ought to be testable since POCOCK & VASANTHY (1986) found that X-ray microanalysis-peaks for silica, aluminum, copper, phosphorus, sulphur, and other elements varied in proportion to their abundance in the soil around the plant producing the pollen. The peaks also varied as a result of conditions of storage and processing of the pollen for X-ray microanalysis. Addition of elements like silicon could, as in doping of components of computer chips, change the "performance" of sporopollenin with regard to the effect of oxidation.

Chemistry of the polymeric substance of exines (sporopollenin) remains controversial (see review by KEDVES 1986). Results of the work of WIERMANN and his group (PRAHL & al. 1986, RITTSCHER & al. 1987, SCHULZE OSTHOFF & WIERMANN 1987) indicate that sporopollenin is a heterogeneous polymer. The design of experiments concerning the chemistry of sporopollenin and the interpretation of results can be aided by our information on the structure of partly degraded exines. We can, for example, estimate the relative volume of the exine components missing or remaining in the 3-D networks, in Fig. 5 the filaments of the net represent an estimated 52% of the relative volume of the exine.

Modeled reconstruction of the exine

A structural model offers a means to in effect make an interpretation with regard to reassembly of exines degraded to the 3-D network level. The illustrations selected for this piece suggest many points requiring consideration in an adequate model, for example:

(1) A network of filaments is a common remnant of partly degraded exines.

(2) Spaces about 40 nm wide are usual in this "fundamental" 3-D network.

(3) Variable amounts of structure remain in these spaces (greatest perhaps in Fig. 3).

(4) Microchannels 20 − 25 nm wide can be tunneled out to more than twice their original diameter.

(5) Cylindrical structures with a core c. 40 nm wide occur on the 3-D net-like background as in Fig. 6.

(6) Laterally oriented white line-centred lamellations up to 200 nm in width may be greatly deflected. They are sometimes clearly evident, sometimes difficult or impossible to detect.

(7) When there are irregular channels wide gaps open up within the endexine, not on white lines but midway between them.

(8) In exines considered to be physiologically inactive there is prominent radially orientated structure across the endexine (as in Fig. 11).

(9) Radial structures in the endexine and tunneled out microchannels both show many short (c. 70 nm wide) lateral components.

(10) In the endoaperture the components I view as exine units have a 40 nm wide uncontrasted cylindrical core zone while the rest of the endexine is "solid" (during physiological activity the c. 40 nm wide holes may again be evident as in Figs. 9 and 10).

(11) The exine is extremely flexible and extensible in height as indicated in Figs. 9 − 11. Change in size during development may involve elongation of unit structures and assimilation of new exine units among preexistent units (ROWLEY & ROWLEY 1986).

The model exine unit that fits these many variations is, I think, cylindrical and in basic form has been referred to as a tuft unit of exine substructure (ROWLEY & DAHL 1977, 1982; ROWLEY & al. 1981 a, b). A tuft in model form consists of a core zone surrounded by a cylindrical "wall" consisting of binder subunits. Core subunits may be straight as in our results with *Nuphar* (FLYNN & ROWLEY 1971) or periodically coiled into loops c. 30 nm in diameter.

Both core and binder subunits are c. 10 nm in diameter. The binders are shown rotary shadowed and negatively stained and the c. 30 nm wide loops of core subunits are given special attention in ROWLEY (1987). The inserted sketch in Fig. 3 shows an interpretation for *Classopollis* that includes such loops in the core region, overlapping or interdigitating with binders (this is the interpretation shown in sketches in ROWLEY & DAHL 1982).

When these (30 nm) loops in core subunits are interdigitated over the central axis the exine appears to be solid. When they are oriented in the "wall" (binder) region rather than directed towards the axis then, as a result, there are c. 40 nm wide holes in the exine. There is one 40 nm wide zone of low contrast in Fig. 10, a few 40 nm wide holes in the endexine and many in the ectexine of Figs. 9 and 12.

Microchannels are usually between 25 and 30 nm in width (AFZELIUS 1956, CHRISTENSEN & al. 1972, DUNBAR & ROWLEY 1984, EL-GHAZALY & JENSEN 1985) whereas exine subunits as we have seen them are about 10 nm in diameter (e.g.,

ABADIE & al. 1986 – 87). I consider it unlikely that microchannels are individual subunits, both because of their size and because in the tunneling out experiments size increase in microchannels was not stepwise, as might be expected if subunits were degraded successively.

My theoretical interpretation is that 30 nm wide loops in the c. 10 nm wide radially arranged subunits swing outward from the core axis leaving a central channel of microchannel dimensions; this is shown in my sketches (ROWLEY 1986: Fig. 13 d; 1988).

From what I have seen white lines probably consist of loops in c. 10 nm wide rods or tubules. Curiously this evidence comes in large part from the green alga *Scenedesmus*. As you know the work with green algae is important just now with regard to accumulation of sporopollenin (ATKINSON & al. 1972). As part of a continuing project concerned with phosphorus metabolism in *Scenedesmus* (e.g., TILLBERG & al. 1984) I have studied the trilamellate (white line-centred) outer sporopollenin layer through many stages in growth. The sporopollenin layer, in my view, originates as an aligned array of c. 30 nm wide loops in a plasma membrane surface coating (glycocalyx) consisting of c. 10nm wide rods or tubules. This same pattern is apparent in discarded sporopollenin lamellations of mother cells after autospore formation. Both the lamellar nature and sporopollenin composition of the white line-centred outer layer is well documented in the work of ATKINSON & al. (1972), BURCZYK & HESSE (1981), BURCZYK & al. (1981).

In systems like *Epilobium* a white line lamellation could consist of a sheet formed of a number of interwoven tubules. These tubules would be composed of c. 10 nm wide subunits (macromolecules) that form ranks of 30 nm loops at the level of the white line. Components of the white line enter at the upper surface and exist below – they do not run parallel with the "lamellation" but, rather, cross it. Examples are seen in the inserts to Fig. 12 where bent arrows indicate subunits that "loop" around white lines and straight arrows indicate profiles of c. 10 nm wide subunits. Several of the most exceptional features of the white line system can be explained by postulating a reversible construction based upon subunits that are uniform throughout and offer a lamellar structure resilient enough to hold together and then reassemble exines during and after irregular channeling. The structurally strong and resilient system could result from an interweaving of the many subunits of each unit structure into different portions of the composite lamellation.

References

ABADIE, M., HIDEUX, M., 1979: L'anthère de *Saxifraga cymbalaria* L. ssp. *huetiana* (BOISS.) ENGL. et IRMSCH. en microscopie électronique (M.E.B. et M.E.T.). 2. Ontogénèse du sporoderme. – Ann. Sci. Nat., ser. Botanique, 13 sér. **1**: 199 – 223.

– – ROWLEY, J. R., 1986 – 87: Ultrastructural cytology of the anther. II. Proposal for a model of exine considering a dynamic connection between cytoskeleton, glycolemma and sporopollenin-synthesis. – Ann. Sci. Nat., ser. Botanique, 13 sér. **8**: 1 – 16.

AFZELIUS, B. M., 1956: Electron micrograph of microchannels in *Corylus*. – In ERDTMAN, G., (Ed.): Current trends in palynological research work. – Grana Palynol. **1**: 127 – 139.

– ERDTMAN, G., SJÖSTRAND, F. S., 1954: On the fine structure of the outer part of the spore wall of *Lycopodium clavatum* as revealed by the electron microscope. – Svensk. Bot. Tidskr. **48**: 155 – 161.

ATKINSON, A. W., Jr., GUNNING, B. E. S., JOHN, P. C. L., 1972: Sporopollenin in the cell wall of *Chlorella* and other algae: ultrastructure, chemistry, and incorporation of 14 C-acetate, studied in synchronous cultures. — Planta **107**: 1 – 32.

AUDRAN, J.-C., 1970: Sur l'ultrastructure de la paroi pollinique chez le *Ceratozamia mexicana* (Cycadacées). — Pollen & Spores **12**: 485 – 511.

— 1977: Recherches cytologiques et cytochimiques sur la genese des grains de pollen et des exines chez les *Cycadales* (Prespermaphytes). — These: Université de Reims.

— 1980: Morphogenèse et altérations provoquées des exines des *Cycadales:* apports à une meilleure interprétation de leur infrastructure. — Rev. Cytol. Biol. Végét., ser. Bot. **3**: 311 – 353.

BURCZYK, J., HESSE, M., 1981: The ultrastructure of the outer cell wall-layer of *Chlorella* mutants with and without sporopollenin. — Pl. Syst. Evol. **138**: 121 – 137.

— SZKAWRAN, H., ZONTEK, I., CZYGAN, F. CH., 1981: Carotenoids in the outer cell-wall layer of *Scenedesmus* (*Chlorophyceae*). — Planta **151**: 247 – 250.

CHRISTENSEN, J. E., HORNER, H. T., Jr., LERSTEN, N. R., 1972: Pollen wall and tapetal orbicular wall development in *Sorghum bicolor* (*Gramineae*). — Amer. J. Bot. **59**: 43 – 58.

CLAUGHER, D., 1986: Pollen wall structure, a new interpretation. — Scanning Electron Microscopy **1**: 291 – 299.

— ROWLEY, J. R., 1987: *Betula* pollen grain substructure revealed by fast atom etching. — Pollen & Spores **29**: 5 – 20.

CRANG, R. E., MAY, G., 1974: Evidence for silicon as a prevalent elemental component in pollen wall structure. — Canad. J. Bot. **52**: 2171 – 2174.

DUNBAR, A., ROWLEY, J. R., 1984: *Betula* pollen development before and after dormancy: exine and intine. — Pollen & Spores **26**: 299 – 338.

EHRLICH, H. G., HALL, J. W., 1959: The ultrastructure of Eocene pollen. — Grana Palynol. **2**: 32 – 35.

EL GHAZALY, G., JENSEN, W. A., 1985: Studies of the development of wheat (*Triticum aestivum*) pollen: 3. Formation of microchannels in the exine. — Pollen & Spores **27**: 5 – 14.

ERDTMAN, G., 1960: The acetolysis technique – a revised description. — Svensk Bot. Tidskr. **54**: 561 – 564.

FLYNN, J. J., ROWLEY, J. R., 1971: Wall microtubules in pollen grains. — Zeiss Information **76**: 40 – 45.

GABARAYEVA, N. I., 1986: The development of the exine in *Michelia fuscata* (*Magnoliaceae*) in connection with the changes in cytoplasmic organelles of microspores and tapetum. — Bot. Ž. **71**: 311 – 322.

— 1987: Ultrastructure and development of sporoderm in *Manglietia tenuipes* (*Magnoliaceae*) during tetrad period: the primexine formation in connection with cytoplasmic organelle activity. — Bot. Ž. **72**: 281 – 290.

GODWIN, H., ECHLIN, P., CHAPMAN, B., 1967: The development of the pollen grain wall in *Ipomoea purpurea* (L.) ROTH. — Rev. Palaeobot. Palynol. **3**: 181 – 195.

HECKMAN, C. A., 1970: Spore wall structure in the *Jungermanniales*. — Grana **10**: 109 – 119.

HIDEUX, M., ABADIE, M., 1985: Cytologie ultrastructurale de l'anthère de *Saxifraga*. 1. Période d'initiation des précurseurs des sporopollénines au niveau des principaux types exiniques. — Canad. J. Bot. **63**: 97 – 112.

KEDVES, M., 1986: In vitro destruction of the exine of recent palynomorphs . — Acta Biol. Szeged. **32**: 49 – 60.

— PÁRDUTZ, A., 1970: Études palynologiques des couches du Tertiaire inférieur de la Région Parisienne. 6. — Pollen & Spores **12**: 553 – 575.

— — 1983: Electron microscope investigations of the early Normapolles pollen genus *Atlantopollis*. — Palynology **7**: 153 – 169.

KERI, CH., 1986: Entwicklungsgeschichte und Ultrastruktur des Sporoderms und der Tapetumzellen von *Clarkia elegans, Epilobium angustifolium* und *Godetia purpurea* (*Onagraceae*). − Diss. Formal. Naturwiss. Fakultät Universität Wien.

KURMANN, M. H., 1986: Pollen wall ultrastructure and development in selected gymnosperms. − Thesis, The Ohio State University.

POCOCK, S. A. J., VASANTHY, G., 1986: Eds analysis of pollen wall surfaces of *Vernonia monosis* CL. (*Asteraceae*) and pollen-soil concentration of elements. − Geophytology **16**: 37−53.

PRAHL, A. K., RITTSCHER, M., WIERMANN, R., 1986: New aspects of sporopollenin biosynthesis. − In MULCAHY, D. L., MULCAHY, G. B., OTTAVIANO, E., (Eds.): Biotechnology and ecology of pollen, pp. 313−318. − Berlin, Heidelberg, New York, Tokyo: Springer.

RITTSCHER, M., GUBATZ, S., WIERMANN, R., 1987: Phenylalanine, a precursor of sporopollenin biosynthesis in *Tulipa* cv. apeldoorn? − Abstracts 14th Internat. Bot. Congress, Berlin, p. 51.

ROWLEY, J. R., 1971: Implications on the nature of sporopollenin based upon pollen development. − In BROOKS, J., GRANT, P. R., MUIR, M., VAN GIJZEL, P., SHAW, B., (Eds.): Sporopollenin, pp. 174−219. − London: Academic Press.

− 1975: Lipopolysaccharide embedded within the exine of pollen grains. − In BAILEY, G. W., (Ed.): 33rd Ann. Proc. Electron Microscopy Soc. Amer., pp. 572−573. − Las Vegas, Nevada.

− 1976: Dynamic changes in pollen wall morphology. − In FERGUSON, I. K., MULLER, J., (Eds.): The evolutionary significance of the exine. − Linn. Soc. Symp. Ser. **1**: 39−66.

− 1978: The origin, ontogeny, and evolution of the exine. − 4th Int. Palynol. Conf., Lucknow (1976−77) **1**: 126−136.

− 1986: A model for plasmodesmata. − In CRESTI, M., DALLAI, R., (Eds.): Biology of reproduction and cell motility in plants and animals, pp. 175−180. − Siena: University of Siena.

− 1987: Plasmodesmata-like processes of tapetal cells. − La Cellule **74**: 229−241.

− 1988: Substructure within the endexine, an interpretation. − J. Palynology (THANIKAIMONI memorial volume) **23/24**: 29−42.

− PRIJANTO, B., 1977: Selective destruction of the exine of pollen grains. − Geophytology **7**: 1−23.

− DAHL, A. O., 1977: Pollen development in *Artemisia vulgaris* with special reference to glycocalyx material. Pollen & Spores **19**: 169−284.

− − 1982: Similar substructure for tapetal surface and exine "tuft"-units. − Pollen & Spores **24**: 5−8.

− ROWLEY, J. S., 1986: Ontogenetic development of microspores of *Ulmus* (*Ulmaceae*). − In BLACKMORE, S., FERGUSON, I. K., (Eds.): Pollen and spores: form and function. − Linn. Soc. Symp. Ser. **12**: 19−33.

− DAHL, A. O., SENGUPTA, S., ROWLEY, J. S., 1981 a: A model of exine substructure based on dissection of pollen and spore exines. − Palynology **5**: 107−152.

− DAHL, A. O., ROWLEY, J. S., 1981 b: Substructure in exines of *Artemisia vulgaris* (*Asteraceae*). − Rev. Paleobot. Palynol. **35**: 1−38.

− SKVARLA, J. J., 1987: Ontogeny of pollen in *Poinciana* (*Leguminosae*). 2. Microspore and pollen grain periods. − Rev. Palaeobot. Palynol. **50**: 313−331.

− SRIVASTAVA, 1986: Fine structure of *Classopollis* exines. − Canad. J. Bot. **64**: 3059−3074.

− EL-GHAZALY, G., ROWLEY, J. S., 1987: Microchannels in the pollen grain exine. − Palynology **11**: 1−21.

SCHULZE OSTHOFF, K., WIERMANN, R., 1987: Phenolics − important constituents of sporopollenin from *Pinus* pollen. − Abstracts 14th Internat. Bot. Congress, p. 52. Berlin.

SKVARLA, J. J., LARSON, D. A., 1965: Interbedded exine components in some *Compositae*. − Southwest. Naturalist. **10**: 65 − 68.

SOUTHWORTH, D., 1985 a: Pollen exine substructure. 1. *Lilium longiflorum*. − Amer. J. Bot. **72**: 1274 − 1283.

− 1985 b: Pollen exine substructure. 2. *Fagus sylvatica*. − Grana **24**: 161 − 166.

− 1986 a: Pollen exine substructure. 3. *Juniperus communis*. − Canad. J. Bot. **64**: 983 − 987.

− 1986 b: Substructural organization of pollen exines. − In BLACKMORE, S., FERGUSON, I. K., (Eds.): Pollen and spores: form and function. − Linn. Soc. Symp. Ser. **12**: 61 − 69.

− SINGH, M. B., HOUGH, T., SMART, I. J., TAYLOR, P., KNOX, R. B., 1988: Antibodies to pollen exines. − Planta **176**: 482 − 487.

TILLBERG, J.-E., BARNARD, T., ROWLEY, J. R., 1984: Phosphorus status and cytoplasmic structures in *Scenedesmus* (*Chlorophyceae*) under different metabolic regimes. − J. Phycol. **20**: 124 − 136.

WAHA, M., 1987: Sporoderm development of pollen tetrads in *Asimina triloba* (*Annonaceae*). − Pollen & Spores **29**: 31 − 44.

Address of author: JOHN R. ROWLEY, Department of Botany, University of Stockholm, S-106 91 Stockholm, Sweden.

Pl. Syst. Evol. [Suppl. 5], 31 – 37 (1990)

The microtubular cytoskeleton during pollen development

S. J. Owens, J. M. Sheldon, and H. G. Dickinson

Received February 12, 1988

Key words: Microsporogenesis, microtubular cytoskeleton, exine formation, exine pattern, intine, tapetum.

Abstract: The focus of this paper is the microtubular cytoskeleton of microspore mother cells and tapetal cells. In both tissues, the microtubular cytoskeleton takes several forms during microsporogenesis and may perform a number of differing roles during meiosis and pollen development. The major part played by microtubules during microsporogenesis is in cell division. In the microspore, the microtubular cytoskeleton, via the microtubular organizing centres (MTOCs), appears to have an indirect role in the siting of the colpus but it plays no direct part in wall patterning. It may also be active in sexine formation in the directional movement of vesicles containing wall precursors, and in the ordered deposition of wall material. Although an association has been observed between microtubules and cisternae forming the nexine 2, it is not known whether these are causally related. It is envisaged that the microtubules play a similar role in intine formation to that played in the development of somatic cellulosic walls.

In amoeboid tapeta, there is increasing evidence to suggest that microtubules may organize cell movement, maintain cell shape, and play a part in exine deposition and, in some cases, in wall patterning.

As more information is accumulated concerning the various constituent parts of the cytoskeleton, these components are being implicated in a number of very significant functions including the directional movement of vesicles (NORTHCOTE 1971, SHEETZ & al. 1986), the maintenance of cell shape (see, e.g., DUSTIN 1978) and, less directly in gene expression (FEY & al. 1986).

In this paper, the occurrence, distribution, and functions of microtubules during microsporogenesis are reviewed with special attention being given to the possible roles of microtubules in pollen wall development. This latter supposition appears logical in that the microtubular cytoskeleton plays an active role in the organization of cell walls in somatic tissues (see, e.g., PALEVITZ 1982), microtubules being implicated in the guiding (LLOYD 1984) and orienting (QUADER 1986) of the cellulose synthase units within the plasma membrane. A similar role appears certain for intine development (DICKINSON & HESLOP-HARRISON 1971).

Microsporogenesis

The microspore mother cell. In premeiotic mitoses, attempts to view microtubular behaviour have failed primarily for technical reasons (HOGAN 1987, SHELDON &

al. 1988). Throughout microsporogenesis itself, microtubules have clearly been shown to be present (SHELDON & DICKINSON 1983, 1986; DICKINSON & SHELDON 1984, 1986; HOGAN 1987) although the microtubular cytoskeleton appears in many forms. In early meiotic prophase, the microtubular cytoskeleton becomes associated with the nuclear envelope, and this organization persists until mid-prophase, and suggests a possible association with the chromosomes. This complex assembly is no longer detectable as the cells enter late prophase and in the nuclear and cell divisions (premeiosis, meiosis and pollen development), microtubules form the spindles and the phragmoplast. Disappointingly, there is no evidence to suggest the presence of pre-prophase bands (see, eg. g., GUNNING & WICK 1985) during meiotic divisions and no unequivocal evidence that the pre-prophase band is present in pollen mitoses (BURGESS 1970, OWENS & WESTMUCKETT 1983, SHELDON & al. 1988).

The tetrad. In the young microspores of the tetrad, microtubules assume a radial organization extending from sites at the nuclear envelope to the inner face of the plasma membrane. In the early tetrad stage vesicles have been observed associated with both this cytoskeleton and the plasma membrane (SHELDON & DICKINSON 1983, 1986). Radially oriented microtubules are present during the development of the sexine (ectexine) and the nexine 2 layers of the pollen wall. When the spore is released from the tetrad and intine synthesis commences, only cortical microtubules can be detected.

The microspore. Microtubules help to maintain the vermiform shape of generative cells. This is achieved by $6-7$, approximately evenly dispersed, microtubular bundles, each containing $6-30$ microtubules, positioned in the cytoplasm close to the generative cell wall, and oriented in the long axis of the cell (SANGER & JACKSON 1971, CRESTI & al. 1984). Microtubules in the bundles of *Nicotiana alata* have recently been shown to have extensive cross bridging which may ensure their structural integrity (LANCELLE & al. 1987). They are not discussed further in this paper.

The tapetum. Tapetal cells of both the parietal and amoeboid type of tapetum lose their pectocellulosic walls during microsporogenesis. Microtubules appear lying adjacent to the plasma membrane of the degrading wall and increase in number during this phase. In the parietal tapetum, the cells remain in situ and the role of cytoplasmic microtubules appears to be to maintain cell shape and position (STEER 1977, PACINI & JUNIPER 1979). In the amoeboid type, differing roles have been proposed for the microtubules depending on the stage of meiosis or pollen development. The activities of the cytoskeleton at each developmental stage are listed below.

Invasion of the loculus. Microtubules may either act to push the tapetal cytoplasm between the meiocytes, or to guide the amoeboid tapetum in the right direction (PACINI & JUNIPER 1983, TIWARI & GUNNING 1986 b).

Tetrad stage. Microtubules may form a type of cage around the tetrads, thereby restraining movement within the loculus (PACINI & JUNIPER 1983).

Microspore stage. The microtubular cage may remain in these later stages (OWENS & DICKINSON 1983, PACINI & JUNIPER 1983). In the fern, *Anemia phyllitidis* (SCHRAUDOLF 1984), and in *Arum italicum* (PACINI & JUNIPER 1983), the microtubules are implicated in the further development of exine pattern; this is discussed later.

The formation of the pollen wall. HESLOP-HARRISON (1968), DOVER (1972), and more recently SHELDON & DICKINSON (1983, 1986) and DICKINSON & SHELDON (1984, 1986) have attempted to determine the role of microtubules in exine formation, and particularly a possible involvement in the development of the pollen wall pattern. This has involved observations of microtubules in situ using the electron microscope and immunocytochemical techniques, by disrupting the polarity of the cell using centrifugation, and by the use of chaotropic agents such as colchicine and cytochalasin.

Determination of pattern. To date there is no experimental evidence to point to the involvement of any specific organelle in the generation of exine pattern. Although microtubules are present at all stages, and their organization has been shown to change quite dramatically during meiosis and pollen development, there is no unequivocal data indicating a definite role in the determination of wall patterning. The fact that cenrifugation treatment during meiotic prophase and at cell plate formation is effective in modifying wall formation indicates that pattern determination may be a lengthy process, only reaching completion in the young tetrad (HESLOP-HARRISON 1971, SHELDON & DICKINSON 1983). DICKINSON & SHELDON (1986) have proposed a model for the generation of the reticulate pattern based on a mechanism of "self assembly" operating at the plasma membrane. The only reports which appear to indicate a direct role for microtubules in wall patterning have been published by SCHRAUDOLF (1984) for the fern, *Anemia phyllitidis*, and by PACINI & JUNIPER (1983) for *Arum italicum*. In the former case, the formation of spore ornamentation takes place during the last phase of exine development. It begins with the appearance of localized assemblies of microtubules at the plasma membrane of the plasmodial tapetum which mark sites of local wall growth. It is unclear whether an influx of precursors is coordinated and directed by the microtubular assemblies or whether the assemblies facilitate a localized retraction of the plasma membrane. In *Arum italicum*, microtubules are associated with a retraction of the plasma membrane.

Position of the colpus. Microtubules, probably in the formation of the spindle, appear to play an important part in the positioning of the colpus (HESLOP-HARRISON 1971, DOVER 1972, SHELDON & DICKINSON 1986). In *Lilium*, if meiosis is disrupted at the prophase stage using colchicine, the colpus may be affected in a variety of ways depending on the stage of division subsequently attained. Cells that remain undivided have either no aperture or a varying number of irregular ones. If only one division takes place the colpus often has a banded appearance. All fully divided pollen grains produce a normal colpus (SHELDON & DICKINSON 1986, SHELDON 1986). This, however, cannot be the whole story since in the early tetrad stage, the apposition of sheets of endoplasmic reticulum to the plasma membrane at the colpal sites shields the cell surface from the wall-generating activities of the young spore cytoplasm.

Centrifugation treatment does not appear to affect the disposition of microtubules within the tetrad cells. It may, however, displace the membranous stencil, and this seems to lead to the formation of "colpal islands" (SHELDON & DICKINSON 1983).

The mechanism by which the spindle controls colpal position is far from clear. It is possible that the spindle MTOCs (microtubule organizing centres) stabilize an area of cytoplasm immediately subjacent to the colpal shield.

Ordered deposition of sexine (ectexine). Microtubules may play a significant part in the transport of exine precursors from the cytoplasm during de-velopment of the wall. Recently, DICKINSON & SHELDON (1984) and SHELDON & DICKINSON (1986) have suggested that the radial arrays of microtubules, which emanate from the MTOCs at the nuclear envelope, may be important in the trans-port of precursors of a second carbohydrate-rich wall which confers height to the sporopollenin containing structures. This conclusion stems from the observation that colchicine applied in the early tetrad stage reduces the height of the exine, giving the muri of the wall a squat appearance.

MEPHAM & LANE (1969), OWENS & DICKINSON (1983), PACINI & JUNIPER (1983), and TIWARI & GUNNING (1986 a, b) all showed microtubules to be present in as-sociation with the membranes of the plasmodial tapetum adjacent to the developing spore wall following release of the pollen from the tetrads. ECHLIN & GODWIN (1968), STEER (1977) and PACINI & JUNIPER (1979) also showed this to be true for species with secretory tapeta. In experiments using colchicine treatment at this stage, TIWARI & GUNNING (1986 a) also found that spore wall deposition was disrupted. There was, however, no evidence from their studies of a relationship between microtubule localization and the development of exine ornamentation.

Formation of the nexine 2 (endexine) and intine. Membranous cis-ternae involved in the formation of the nexine 2 layer move to the protoplast surface late in the tetrad stage, and cortical microtubules may regularly be seen associated with these cisternae (DICKINSON & HESLOP-HARRISON 1971). Microtubular function in this process remains unknown. Sporopollenin is then observed to be deposited on lamellae at the outer face of the plasma membrane before apparent transport through the developing intine layer to the developing nexine 2 (ROWLEY & DUNBAR 1967, GODWIN & al. 1967, HESLOP-HARRISON 1968 a). It has, however, been pro-posed that the impression given of the nexine 2 elements travelling through the intine is misleading, and that the two walls are synthesized simultaneously, only to be compacted on vacuolation of the young spore (DICKINSON 1976).

Since the intine layer is now widely-held to be cellulosic in nature, the presence of cortical microtubules subjacent to the plasma membrane during the formation of this layer is not surprising.

Conspectus

Investigations into the structure and function of cytoskeletal components during meiosis and pollen development have largely been restricted to microtubules, the subject of this paper, and it would appear that they perform significant and varied roles during pollen development. There is still very little data on other elements of the cytoskeleton, in particular actin microfilaments and intermediate filaments, which may play a part either individually or in combination with other elements during microsporogenesis. To date there is no evidence for the occurrence of in-termediate filaments in pollen, although there are indications of their presence in plant cells (DAWSON & al. 1985), but there are increasing numbers of reports of actin microfilaments (SCHMIT & LAMBERT 1985, HEPLER & LANCELLE 1986, HES-LOP-HARRISON & al. 1986, SHELDON & HAWES 1988, see also review of STAIGER & SCHLIWA 1987).

As well as their role in nuclear division, microtubules may be involved in chro-mosome pairing (SHELDON & al. 1988), and they appear to be involved in exine production but not the generation of exine pattern. Actin microfilaments have been implicated in cell wall deposition in somatic cells, largely because of their close association with cortical microtubules (see review of HEATH & SEAGULL 1982, TRAAS & al. 1987) and thus require examination for a similar role in developing meiocytes. Unfortunately, work by SHELDON & HAWES (1988) shows that at no time during meiosis do the microfilaments take up a pattern that could be held responsible for the development of the exine pattern in *Lilium*. Actin could, however, still be involved in pollen wall formation. From experiments using cytochalasin B, QUADER & SCHNEPF (1986) have suggested that actin is involved in the distribution of endoplasmic reticulum (ER). This is interesting when considering pollen wall patterning since ER is involved in determining apertures and, in some species, has been implicated in the control of exine deposition (HESLOP-HARRISON 1963, 1968 b; DICKINSON 1970).

References

BURGESS, J., 1970: Microtubules and cell division in the microspore of *Dactylorchis fuschii*. − Protoplasma **69**: 253−264.

CRESTI, M., CIAMPOLINI, F., KAPIL, R. N., 1984: Generative cells of some angiosperms with particular emphasis on their microtubules. − J. Submicrosc. Cytol. **16**: 317−326.

DAWSON, P., HULME, J., LLOYD, C. W., 1985: Monoclonal antibody to intermediate filaments antigen cross reacts with higher plant cells. − J. Cell Biol. **100**: 1793−1798.

DICKINSON, H. G., 1970: Ultrastructural aspects of primexine formation in the microspore tetrad of *Lilium longiflorum*. − Cytobiologie **1**: 437−449.

− 1976: Common factors in exine deposition. − In FERGUSON, I. K., MULLER, J., (Eds.): The evolutionary significance of the exine. − Linn. Symp. Soc. Ser. **1**: 67−89.

− HESLOP-HARRISON, J., 1971: The mode of growth of the inner layer of the pollen grain exine in *Lilium*. − Cytobios **4**: 233−243.

− SHELDON, J. M., 1984: A radial system of microtubules extending between the nuclear envelope and the plasma membrane during early male haplophase in flowering plants. − Planta **161**: 86−90.

− − 1986: The generation of patterning at the plasma membrane of the young microspore of *Lilium*. − In BLACKMORE, S., FERGUSON, I. K., (Eds.): Pollen and spores: form and function. − J. Linn. Symp. Soc. Ser. **12**: 1−17.

DOVER, G. A., 1972: The organisation and polarity of pollen mother cells of *Triticum aestivum*. − J. Cell Sci. **11**: 699−711.

DUSTIN, P., 1978: Microtubules. − Berlin, Heidelberg, New York: Springer.

ECHLIN, P., GODWIN, H., 1968: The ultrastructure and ontogeny of pollen in *Helleborus foetidus* L. 1. The development of the tapetum and Ubisch bodies. − J. Cell Sc. **3**: 161−174.

FEY, E. G., ORNELLES, D. A., PENMAN, S., 1986: Association of RNA with the cytoskeleton and the nuclear matrix. − In LLOYD, C. W., HYAMS, J. S., WARN, R. M., (Eds.): The cytoskeleton: cell function and organization. − J. Cell Sci. Suppl. **5**: 99−119.

GODWIN, H., ECHLIN, P., CHAPMAN, B., 1967: The development of the pollen grain wall in *Ipomoea purpurea* (L.) ROTH. − Rev. Palaeobot. Palynol. **3**: 181−195.

GUNNING, B. S. E., WICK, S. M., 1985: Preprophase bands, phragmoplasts and spatial control of cytokinesis. − J. Cell Sci. Suppl. **2**: 157−179.

HEATH, I. B., SEAGULL, R. W., 1982: Oriented cellulose fibrils and the cytoskeleton: a critical comparison of models. − In LLOYD, C. W., (Ed.): The cytoskeleton in plant growth and development, pp. 165−187. − London: Academic Press.

HEPLER, P. K., LANCELLE, S. A., 1986: Cytoskeletal details in freeze-substituted meiotic cells of *Tradescantia blossfeldiana*. − J. Cell Biol. **103**: 555 a.

HESLOP-HARRISON, J., 1963: An ultrastructural study of pollen wall ontogeny in *Silene pendula*. — Grana Palynol. **4**: 7 – 24.
— 1968 a: Pollen wall development. — Science **161**: 230 – 237.
— 1968 b: Wall development within the microspore tetrad of *Lilium longiflorum*. — Canad. J. Bot. **46**: 1185 – 1192.
— 1971: Wall pattern formation in angiosperm microsporogenesis. — In: Control mechanisms of growth and differentiation. — Symp. Soc. Exper. Biol. **25**: 277 – 300.
— HESLOP-HARRISON, Y., CRESTI, M., TIEZZI, A., CIAMPOLINI, F., 1986: Actin during pollen germination. — J. Cell Sci. **86**: 1 – 8.
HOGAN, C. J., 1987: Microtubule patterns during meiosis in two higher plant species. — Protoplasma **138**: 126 – 136.
LANCELLE, S. A., CRESTI, M., HEPLER, P. K., 1987: Ultrastructure of the cytoskeleton in freeze-substituted pollen tubes of *Nicotiana alata*. — Protoplasma **140**: 141 – 150.
LLOYD, C. W., 1984: Toward a dynamic helical model for the influence of microtubules in wall patterns in plants. — Internat. Rev. Cytol. **86**: 1 – 51.
MEPHAM, R. H., LANE, G., 1969: Formation and development of tapetal periplasmodium in *Tradescantia bracteata*. — Protoplasma **68**: 175 – 192.
NORTHCOTE, D. H., 1971: Organization of structure, synthesis and transport within the plant during cell division and growth. — Symp. Soc. Exper. Biol. **25**: 51 – 69.
OWENS, S. J., DICKINSON, H. G., 1983: Pollen wall development in *Gibasis* (*Commelinaceae*). — Ann. Bot. **51**: 1 – 15.
— WESTMUCKETT, A. D., 1983: The structure and development of the generative cell wall in *Gibasis karwinskyana*, *G. venustula* and *Tradescantia blossfeldiana* (Commelinaceae). — In MULCAHY, D. L., OTTAVIANO, E., (Eds.): Pollen: Biology and implications for plant breeding, pp. 149 – 157. — New York: Elsevier Biomedical.
PACINI, E., JUNIPER, B. E., 1979: The structure of pollen grain development in the Olive (*Olea europaea*). 2. Secretion by the tapetal cells. — New Phytol. **83**: 165 – 174.
— — 1983: The ultrastructure of the formation and development of the amoeboid tapetum in *Arum italicum* MILLER. — Protoplasma **117**: 116 – 129.
PALEVITZ, B. A., 1982: The stomatal complex as a model of cytoskeletal participation in cell differentiation. — In LLOYD, C. W., (Ed.): The cytoskeleton in plant growth and development, pp. 345 – 376. — London: Academic Press.
QUADER, H., 1986: Cellulose microfibril orientation in *Oocystis solitai:* proof that microtubules control the alignment of the terminal complexes. — J. Cell Sci. **83**: 223 – 234.
— SCHNEPF, E., 1986: Endoplasmic reticulum and cytoplasmic streaming: fluorescence microscopical observations in adaxial epidermis cells of onion bulb scales. — Protoplasma **131**: 250 – 252.
ROWLEY, J., DUNBAR, A., 1967: Sources of membranes for exine formation. — Svensk Bot. Tidskr. **61**: 49 – 64.
SANGER, R., JACKSON, W. T., 1971: Fine structure of pollen development in *Haemanthus katharinae* BAKER. 2. Microtubules and elongation of generative cells. — J. Cell Sci. **8**: 303 – 305.
SCHMIT, A. C., LAMBERT, A. M., 1985: F-actin distribution during the cell cycle of higher plant endosperm cells. — J. Cell Biol. **101**: 38 a.
SCHRAUDOLF, H., 1984: Ultrastructural events during sporogenesis of *Anemia phyllitidis* (L.) Sw. 2 Spore wall formation. — Beiträge Biol. Pflanzen **59**: 237 – 260.
SHEETZ, M. P., VALE, R., SCHNAPP, B., SCHROER, T., REESE, T., 1986: Vesicle movements and microtubule-based motors. — In LLOYD, C. W., HYAMS, J. S., WARN, R. M., (Eds.): The cytoskeleton: cell function and organization. — J. Cell Sci. Suppl. **5**: 181 – 188.
SHELDON, J. M., 1986: The generation of pattern in the pollen wall of *Lilium*. — Ph.D. Thesis, University of Reading.

— DICKINSON, H. G., 1983: Determination of patterning in the pollen wall of *Lilium henryi*.
   — J. Cell Sci. **63**: 191—208.
— — 1986: Pollen wall formation in *Lilium:* the effect of chaotropic agents, and the
   organization of the microtubular cytoskeleton during pattern development. — Planta
   168: 11—23.
— HAWES, C. R., 1988: The actin cytoskeleton during male meiosis in *Lilium*. — Cell Biol.
   Internat. Reports **12**: 471—476.
— WILLSON, C., DICKINSON, H. G., 1988: Interaction between the nucleus and cytoskeleton
   during the pairing stage of male meiosis in flowering plants. — In BRANDHAM, P. E.,
   (Ed.): Kew Chromosome Conference **3**: 27—35. — London, Kew: Her Majesty's Sta-
   tionary Office.
STAIGER, C. J., SCHLIWA, M., 1987: Actin localization and function in higher plants. —
   Protoplasma **141**: 1—12.
STEER, M. W., 1977: Differentiation of the tapetum in *Avena*. 1. The cell surface. — J.
   Cell Sci. **25**: 125—138.
TIWARI, S. C., GUNNING, B. E. S., 1986a: An ultrastructural, cytochemical and immu-
   nofluorescence study of post-meiotic development of plasmodial tapetum in *Tradescantia virginiana* L. and its relevance to the pathway of sporopollenin secretion. —
   Protoplasma **133**: 100—114.
— — 1986b: Colchicine inhibits plasmodium formation and disrupts pathways of sporopollenin secretion in the anther tapetum of *Tradescantia virginiana* L. — Protoplasma
   133: 115—128.
TRAAS, J. A., DOONAN, J. H., RAWLINS, D. J., SHAW, P. J., WATTS, J., LLOYD, C. W.,
   1987: An actin network is present in the cytoplasm throughout the cell cycle of carrot
   cells and associated with the dividing nucleus. — J. Cell Biol. **105**: 387—395.

Addresses of the authors: S. J. OWENS, Jodrell Laboratory, Royal Botanic Gardens,
Kew, Richmond, Surrey TW9 3DS, England. — J. M. SHELDON, Department of Plant
Sciences, South Parks Road, Oxford, England. — H. G. DICKINSON, Plant Science Lab-
oratories, The University of Reading, Whiteknights, Reading, England.

Pl. Syst. Evol. [Suppl. 5], 39–51 (1990)

The pollen surface in wind-pollination with emphasis on the *Compositae*

Margaret R. Bolick

Received December 6, 1987; in revised form March 14, 1988

Key words: Angiosperms, *Compositae.* – Palynology, wind-pollination, functional morphology, fluid mechanics.

Abstract: Theoretical calculations and experimental observations indicate that the surface sculpture of pollen has a slight influence on pollen flight dynamics in wind-pollinated angiosperms. Sculpture is also important as it influences pollen clumping. Variations in ultrastructure that affect pollen density will change flight parameters. Wodehouse's hypothesis that the smoothness of the pollen surface in anemophilous plants results from an overall thinning of the exine is not supported by measurements of the exine in the *Compositae* or for angiosperms as a whole.

The pollen of anemophilous plants is smooth when compared to the often highly ornamented pollen of entomophilous plants. The question of why this is so has been answered with several theories in the past 50 years. These theories may be grouped into two categories: (a) explanations that regard smooth surface sculpture as a secondary effect of another phenomena such as exine thinning; and (b) explanations that propose adaptationist scenarios for surface sculpture patterns. This paper reviews, within the framework of fluid mechanics, the evidence supporting the different hypotheses and presents new observations testing the alternatives.

Wodehouse, in his book "Pollen Grains" (1935), was among the first to propose an hypothesis that viewed surface sculpture as a secondary effect of other phenomena. Working specifically on the *Ambrosiinae* and *Artemisia* (*Compositae*) he said that the exine of pollen transported by the wind was thinner than the exine of pollen transported by animals. He thought that the thinning of the exine results in loss of sculpture.

Wodehouse's hypothesis has not been challenged by workers such as Faegri & Iversen (1975: 52), Lewis (1977), and Whitehead (1969, 1983) although Lee (1978) showed that pollen exine thickness is scaled to pollen diameter.

Faegri & van der Pijl (1979: 38) are among those suggesting a slightly different emphasis, stating that elaborate surface sculpture facilitates animal pollination (see also Proctor & Yeo 1972: 258, Frankel & Galun 1977: 35).

Whitehead (1969, 1983) and others mentioned in passing that smoothness must somehow be better for pollen flight or capture. In an apparent contradiction to

WHITEHEAD, SCHEPPEGRELL (1917) and COCKE (1937) proposed that the small spines on ragweed (*Ambrosia*) pollen grains make them fall more slowly than would comparably sized completely smooth pollen grains. GREGORY (1973) and NIKLAS (1985 b), while not speaking specifically of *Compositae* pollen, noted that surface roughness should decrease pollen terminal velocity.

In sharp contrast, HUTCHINSON (1967) stated that surface roughness per se is unlikely to retard settling speed at low Reynolds numbers. CRANE (1986), also considered the large boundary layers present at low Reynolds numbers and suggested that the absence of pronounced sculpture on the pollen surface in wind-dispersed angiosperm pollen might be related to avoidance of secondary relofting of pollen from stigmatic surfaces.

LEE's (1978) work showing a correlation between pollen diameter and exine thickness raised the possibility that the apparent thinness of the exine in wind-pollinated plants may be only an artefact of the generally smaller size range of pollen found in these plants. If pollen diameter and exine thickness are proportionately scaled, then animal vectored pollen, since it is on the average larger in diameter, would have, again on the average, thicker walls[1]. This leads to an alternative to WODEHOUSE's (1935) hypothesis that exines are thinner and therefore smoother in anemophilous plants, an alternative that states that the exine of pollen transported by the wind is not thinner than the exine of pollen transported by animals if pollen of the same diameter is compared. Comparative measurements of pollen diameter, exine thickness, and their ratio will be presented to evaluate these theories.

The second group of hypotheses to be investigated further are those of SCHEPPEGRELL, COCKE, GREGORY, and NIKLAS, relating the pollen surface to some functional aspects of flight dynamics. These will be evaluated for their congruence with fluid mechanics both theoretically and using experimentally determined data on pollen size, density, and retention by stigmas.

Materials and methods

Measurements to evaluate WODEHOUSE's hypothesis were taken from three sources. The first set of 69 measurements of *Compositae* pollen diameter and exine thickness were taken or made from published transmission electron micrographs (TEM) of PAYNE & SKVARLA (1970), PRAGLOWSKI & GRAFSTRÖM (1980), SKVARLA & LARSON (1965), and SKVARLA & al. (1978 b). A second, much larger, set of measurements of pollen of 917 species from the *Compositae* was made using a Leitz Ortholux microscope, an oil immersion objective, and prepared slides in D. A. LIVINGSTONE's pollen reference collection, housed in the Department of Zoology, Duke University. Pollen diameter in species with an echinate surface was measured using the point where the bases of the spines were contiguous in optical cross section as the outer edge. The final set of 91 measurements of various angiosperms was taken from LEE (1978).

Pollen diameter/exine thickness ratios were compared for the wind and animal pollinated species. Linear regressions were also calculated for the pollen diameters and exine thicknesses and their slopes were tested for equality following SOKAL & ROHLF (1981: chapter 14).

HUTCHINSON (1967) and VOGEL (1981) were the major references for fluid dynamics

---

[1] The discussion here and in succeeding paragraphs assumes an average over all of the angiosperms. There are, no doubt, specific cases where a particular anemophilous plant may have pollen with a thinner exine than that found in the pollen of a zoophilous relative.

and physical parameters such as the density of air. Pollen density figures were obtained from DURHAM (1946) and HARRINGTON & METZGER (1963); the pollen terminal velocities of DURHAM (1946) were used. Sporopollenin densities were taken from SOUTHWORTH (1988). Precise morphological measurements on *Ambrosia* pollen were taken from ROBBINS & al. (1979).

Experiments to determine if pollen surface sculpture affected retention on the stigma were conducted in the Duke University Department of Zoology's large wind tunnel. Stigmas of ragweed were measured from fresh and herbarium materials to determine the size of the models. The stigmas were modeled using cylinders of the same diameter and length made of human hair or plastic brush bristles. Pollen of *Ambrosia*, *Eupatorium*, *Artemisia*, and *Chrysanthemum* was obtained from Greer Laboratories, Lenoir, NC. Wind speeds typical for plant canopies were estimated from figures given in WHITEHEAD (1983).

The artificial stigmas were glued to the ends of a toothpick stem. The artificial stigma was coated with pollen by dipping it in the vial of dry pollen from Greer Laboratories. The model stigma was then checked under × 70 magnification and if there was an even coat of pollen, the whole model was placed in a clamp mounted on a post in the center of the wind tunnel. Windspeeds up to 20 m/s for times up to 5 min were used in attempts to see if pollen could be dislodged. The stigmas models were reexamined under × 70 magnification after each exposure to the wind tunnel. In these trials, *Artemisia* and *Chrysanthemum* were always paired as were *Ambrosia* and *Eupatorium*; this gave one long-spined and one short-spined pollen type per trial while controlling for pollen size and ultrastructural type.

Results

Exine thickness. Table 1 shows the average exine thickness of pollen, in μm, of *Ambrosia*, of other taxa in the *Ambrosiinae*, and of animal vectored *Heliantheae* using the measurements from TEM. The average exine thickness for the *Ambrosiinae* is greater than that of the animal-pollinated *Heliantheae*. Table 2, also using TEM measurements, extends the comparison to the other tribes in the *Compositae* with the helianthoid exine morphology. Only the *Calenduleae* has pollen with exines that are thicker than those of the wind-pollinated *Ambrosiinae*.

The exines of the pollen of the *Ambrosiinae* may not be thinner than their animal-pollinated counter parts but these exines are proportioned differently (Table 3). If one looks at the partitioning between the upper tectum (above the cavus) and the footlayer-endexine below, one can see that the footlayer-endexine in the *Ambrosiinae* pollen grain is unusually thin giving this part of the exine greater flexibility to form the air bladders that lower the effective density of the pollen.

With the larger sample size provided by the light microscope measurements

Table 1. Exine thickness in wind- and animal-pollinated *Heliantheae* (*Compositae*). Measurements taken from published TEM studies

Taxon	Thickness (μm)
Wind-pollinated *Heliantheae*	
Ambrosia	1.77
Other *Ambrosiinae*	1.90
All *Ambrosiinae*	1.82
Animal-pollinated *Heliantheae*	1.74

Table 2. Exine thickness in helianthoid exine types from the *Compositae*. Measurements taken from published TEM studies

Taxon	Thickness (μm)
Astereae	0.86
Eupatorieae	1.21
Inuleae	1.66
Helenieae	1.70
Animal-pollinated *Heliantheae*	1.74
Senecionieae	1.76
Wind-pollinated *Heliantheae*	1.82
Calenduleae	1.95
All helianthoid exine types	1.63

Table 3. Exine partitioning in helianthoid pollen

Taxon	Upper tectum (μm)	Footlayer-endexine (μm)	T/F ratio
Ambrosiinae	1.44	0.38	0.26
Heliantheae (animal-pollinated)	1.02	0.72	0.70
Astereae	0.64	0.22	0.34
Eupatorieae	0.89	0.32	0.36
Inuleae	1.20	0.46	0.38
Helenieae	1.16	0.54	0.46
Senecioineae	1.14	0.62	0.54

(Table 4), comparison of pollen diameter/exine thickness ratios shows that the exine of pollen in the *Ambrosiinae* is proportionately thicker than the exine of the comparably sized animal-pollinated *Heliantheae*. Comparing regressions for the *Ambrosiinae* and the rest of the *Heliantheae* (Table 5) results in the rejection of the null hypothesis that the samples are from the same population at the 95% level but not the 99% level. In the *Anthemideae*, the exine of *Artemisia* pollen is slightly thinner than the exine of entomophilous members of the tribe when actual measurements or ratios are compared. When regressions are compared, however, the null hypothesis that the regressions are from the same population is not rejected.

In the 91 species from Lee's data, again, the wind pollinated species do not have thicker exines if pollen diameter/exine thickness ratios are compared (Table 4). Similarly, the regressions are not statistically different (Table 5).

Comparing the regressions of the four *Compositae* samples or the three samples from Lee's data shows that the null hypothesis, that all the regressions are from the same population, cannot be rejected for either.

Fluid mechanics. Durham's (1946) terminal velocities were used as the best estimates available; there is reason to believe that there may be serious flaws in the earlier estimates of Bodmer (1922), Dyakowska (1937), and Knoll (1932)

Table 4. Exine thickness in wind- and animal-pollinated plants. [a] From LEE (1978)

Group	Pollen diameter (in μm)	Exine thickness (in μm)	D/T ratio
Heliantheae			
Wind	22.5 ± 3.5	3.0 ± 0.8	7.5
Animal	29.5 ± 7.4	3.0 ± 0.8	9.8
Anthemideae			
Wind	23.8 ± 4.0	3.6 ± 0.8	6.6
Animal	28.7 ± 4.2	4.6 ± 0.9	6.2
Angiosperms[a]			
Wind	26.0 ± 9.8	2.0 ± 1.1	13.0
Insect	37.5 ± 18.0	2.5 ± 1.9	15.0
Vertebrate	60.0 ± 23.5	3.6 ± 3.1	16.7

Table 5. Comparisons of regressions of pollen diameter (x) and exine thickness (y). [a] From LEE (1978); [b] significant at the 95% level

Group	Regression equation	F
Anthemideae		
Wind	$y = 0.85 + 0.120\,x$	$F_{(1, 88)}$ wind-animal
Animal	$y = 2.66 + 0.077\,x$	$= 0.863$
Heliantheae		
Wind	$y = -0.23 + 0.143\,x$	$F_{(1, 68)}$ wind-animal
Animal	$y = 1.43 + 0.060\,x$	$= 4.341$[b]
		$F_{(3, 156)}$ all *Compositae*
		$= 2.466$
Angiosperms[a]		
Wind	$y = -0.22 + 0.086\,x$	$F_{(1, 76)}$ wind-insect
Insect	$y = 0.10 + 0.065\,x$	$= 0.468$
Vertebrate	$y = -2.30 + 0.099\,x$	$F_{(1, 70)}$ insect-vertebrate
		$= 2.733$
		$F_{(2, 91)}$ all angiosperms
		$= 1.666$

that have been repeatedly cited in more recent works (GREGORY 1973: 22). In the 1946 paper, DURHAM discussed the methods used to determine pollen terminal velocities and the difficulties that he encountered in their estimation. He noted that experimental determinations of pollen terminal velocities differed if the tubes in which the pollen was dropped did not have the same diameter or if the amount of pollen dropped was not the same.

Although DURHAM did not know why the diameter of the tube in which the pollen is dropped affected the terminal velocity, this can now be explained as a wall effect. This phenomenon is most critical at low Reynolds numbers (Re) (VOGEL

1981: 249). Wall effects increase drag and thus would lower the experimentally determined terminal velocity. Vogel (1981: 250) gave a formula from White for estimating when wall effects will become negligible. Calculating Reynolds numbers for a range of pollen types (Re 0.02 – 0.06) and using White's formula indicates that the minimum distance from the falling pollen to the nearest wall should be in the range of 12 to 15 mm, giving a tube diameter of from 24 to 30 mm if one could be sure that all of the pollen would fall in the center of the tube. As Durham said that Bodmer's glass tube was 35 mm in diameter and that Knoll's was similar, their terminal velocities may be too slow as a result of wall effects. As Durham's tube was 51 mm in diameter and he only recorded the fall of grains in the center square centimeter, his results are less likely to be biased by this.

Durham also noted that errors in estimating the terminal velocity would result if too much pollen was dropped at one time; Gregory (1973: 25) discussed this as a bulk sedimentation effect. Niklas (1984) also noted problems caused by pollen clumping and indicated that this could increase the estimated terminal velocity by a factor of 3.3. (See Niklas 1984 for a modern technique for measuring terminal velocity that avoids these problems.)

The critical feature in flight dynamics is the terminal velocity or settling speed. This determines how far the pollen will travel before gravity brings it to the ground. To understand how terminal velocity is affected by surface features, however, one must first know whether inertial or viscous forces dominate; the Reynolds number (Re) provides an estimate of this (Vogel 1981: 65). At high Reynolds numbers, inertial forces dominate; at low Reynolds numbers, viscous forces dominate (Vogel 1981: chap. 13).

The Reynolds number is dimensionless and is calculated by multiplying the length of the object and its velocity and dividing the product by the fluid kinematic viscosity (Vogel 1981). Comparability of Reynolds number implies the same fluid dynamic behavior (Vogel 1981: 68).

The Reynolds number for *Ambrosia* (ragweed) can be calculated (length = diameter = 20 μm for ragweed, Durham 1946; fluid kinematic viscosity of air at 20 °C = 15.00×10^{-6} m^2/s, Vogel 1981: 17) and is in the very low range, 0.01 to 0.02, depending on the terminal velocity used (0.0088 m/s, Durham 1946; 0.0156 m/s, Raynor & al. 1970). Reynolds numbers for pollen of other anemophilous species, calculated from the figures in Durham (1946) are generally in the range 0.02 – 0.06; the major exceptions with much higher Re are cultivated cereals (*Zea, Secale*). These low Reynolds numbers for ragweed and other pollen indicate two things. First, they show that Stokes' law for the drag on a sphere, which holds for Re under 0.5, applies in almost all cases; and when combined within a calculation of the gravitational pull on a grain will give an equation to predict the terminal velocity of ragweed pollen. This prediction should agree with experimental results if surface roughness is unimportant. Secondly, it means that Hutchinson's equations for sinking diatoms can be used for pollen as the Reynolds numbers are comparable (Hutchinson's 1967 Re for a diatom is 0.025).

The equation for Stokes' law is:

$$D = 6\pi u r \hat{v} \qquad (1)$$

where $\hat{v}$ is the terminal velocity, r is the radius of the falling sphere and u is the fluid dynamic viscosity. To solve for $\hat{v}$, one assumes that at a constant rate of descent, drag and acceleration due to gravity are equal. This gives the following equation (from HUTCHINSON 1967, VOGEL 1981):

$$\hat{v} = 2/9\,g\,r^2\,(p' - p)\,u^{-1} \tag{2}$$

where g is the acceleration due to gravity and $p' - p$ is the difference in density between the fluid and the object, in this case, pollen and air.

As can be seen from Eq. (2), increasing the radius will increase the terminal velocity unless the added material is less dense than the original. (This is why the air bladders in ragweed are so effective; the pollen density is decreased without changing the radius.) HUTCHINSON (1967: 274) also gave an equation to show how much less dense the added material must be in order to decrease terminal velocity. He considered adding extra material in the form of a gelatinous sheath to the outside of a diatom and gave the following equation to show its effect on diatom terminal velocity:

$$\hat{v} = 2/9\,g\,r^2\,\hat{a}^2\,[p'' - p + 1/\hat{a}^3\,(p' - p'')]\,u^{-1} \tag{3}$$

where $\hat{a}$ is the proportion by which the radius is increased and p'' is the density of the added material. Rearranging terms, HUTCHINSON derives the following inequality:

$$\frac{p' - p''}{p'' - p} > \hat{a}\,(\hat{a} + 1) \tag{4}$$

This shows that if the difference in density between the object and the added material ($p' - p''$) is not greater than twice the density difference between the added material and the medium ($p'' - p$), the added material will not decrease terminal velocity. The density of air at 20 °C, $1.205\,kg/m^3$, is so much less than the densities typical of pollen, $1\,000\,kg/m^3$, (approx. that of water, CRANE 1986) that this part of the equation can be ignored. Reworking the inequality for pollen in light of this, the effective density of the added material must be less than one-third that of the grain. Since the range of experimentally determined densities of sporopollenin is 1.1 to 1.4 times that of water (or the whole grain) (SOUTHWORTH 1988), the added material must incorporate 70 to 77% air.

The shape of spines on ragweed pollen is nearly conical and thus their volume can be approximated using the geometrical formula for volumes of this shape. Doing this from measurements of spine height, determining the proportional base width and multiplying by spine number using the figures in ROBBINS & al. (1979), allows one to calculate the proportions of air and sporopollenin incorporated in the volume resulting from a given increase in pollen radius (Table 6). As the percentage of air in the total volume increase is below 0.70 for both species, spines will not decrease the terminal velocity for ragweed pollen.

Values of $\hat{a}$ can be calculated from ROBBINS & al. (1979) (Table 6) and used to estimate the effects of spines in increasing terminal velocity. Approximate densities of added material can be calculated from the spine proportion times 1.25 (the midpoint of the range of sporopollenin densities; SOUTHWORTH 1988). Using these values in Eq. 3 shows that adding spines increases the terminal velocity of the longer-spined *A. trifida* L. by a factor of 1.2 and of the shorter-spined *A. artemisiifolia* L. by a factor of 1.1. To put this in perspective, the variations in ragweed pollen density, noted by HARRINGTON & METZGER (1963), resulting from differences

M. R. Bolick:

Table 6. Spine characteristics and flight parameters for *Ambrosia trifida* and *A. artemisiifolia*. Measurements from Robbins & al. (1979) except where noted

Species	Radius with spines[a]	Spine			
		height	base[b]	number	volume[c]
A. trifida	9.71 μm	1.80 μm	2.4 μm	71.42	775 μm³
A. artemisiifolia	9.59 μm	1.39 μm	2.1 μm	87.31	564 μm³

Spines, % added volume[d]	â[e]	Density, added volume[f]	v̂		
			with spines[g]	without spines[h]	exper.[i]
43	1.23	0.54 g/cm³	0.96 cm/s	0.76 cm/s	0.82 cm/s
40	1.17	0.40 g/cm³	0.94 cm/s	0.75 cm/s	0.88 cm/s

[a] Figures used here are the average of the polar and equatorial diameters
[b] Estimated from spine base to height ratio in SEM pictures of Robbins & al. (1979) and their figures for spine height
[c] Calculated from spine height, the estimate of spine base length, the formula for the volume of a cone, and the number of spines per grain
[d] The increase in volume resulting from an increase in radius, equal to the spine length, was calculated. The spine volume was taken as a percentage of this
[e] â is the proportional increase in radius due to added material used in Eq. (3). In this case it is due to spine length
[f] The density of the added volume was calculated by multiplying the percentage of the added volume that was spines times 1.25 g/cm³, the midpoint of the range of densities of sporopollenin (Southworth 1988)
[g] Terminal velocity (v̂) was predicted using Eq. (2), based on Stokes' law, and using a pollen radius that included the full spine length
[h] Terminal velocity (v̂) was predicted using Eq. (3) and including spine length under â, rather than as part of the radius
[i] Terminal velocity (v̂) determined experimentally by Durham 1946

in the relative humidity of the air around the pollen, also produce a range of values for terminal velocity. The most humid, most dense ragweed pollen has a terminal velocity that is 1.5 times that of the driest, least dense pollen.

This leads back to experiments of Durham (1946) on ragweed pollen density and terminal velocity. Durham underestimated the pollen density (his estimate was 0.55 g/cm³) and thus he believed his experimentally determined terminal velocities for ragweed did not follow Stokes' law. The terminal velocity can be recalculated using the correct pollen density for dried ragweed pollen, as determined by Harrington & Metzger (1963) (0.84 g/cm³), and pollen radii, as determined by Robbins & al. (1979). (Durham's measurements of diameter are less appropriate as they were made on wet, expanded grains.) This was done in two ways. The terminal velocities for *A. trifida* and *A. artemisiifolia* were first calculated using

Eq. 2 and a pollen radius that included spine length. The terminal velocities were recalculated using Eq. 3, a pollen radius that excluded spine length, a value for â that incorporated the spine length, and a density for the added spines determined by the procedures discussed earlier. The two sets of predictions bracket DURHAM's experimental measurements (Table 6) and are closer to them than the predictions he made.

Experiments on stigmatic retention of pollen. In many trials, with wind speeds up to gale force, there was never enough relofting of pollen to produce a detectable loss when the stigmas were reexamined.

Discussion

CRANE (1986) noted that wind-pollination has not received as much attention from researchers as has animal-pollination. This is changing as CRANE's work on pollen function (1986) and NIKLAS' work on the aerodynamics of pollen transport and capture (CAMAZINE & NIKLAS 1984; NIKLAS 1981 a, b, 1982, 1983, 1984, 1985 a, b, c, 1987; NIKLAS & BUCHMANN 1985, 1987; NIKLAS & al. 1986; NIKLAS & KERCHNER 1986; NIKLAS & NORSTOG 1984; NIKLAS & PAW U. 1982, 1983) have increased understanding of this mode of abiotic pollination.

The measurements of pollen size and exine thickness do not support WODE-HOUSE's hypothesis that anemophilous pollen has a thinner exine. In the groups investigated, the regression of pollen diameter and exine thickness was significantly different only where the wind-pollinated plants had a proportionately much thicker exine, in the *Ambrosiinae* in comparison with the *Heliantheae*, and only at the 95% confidence level. For the other comparisons, wind pollinated plants did not have pollen with proportionately thinner exines. These results cast serious doubt on WODEHOUSE's theory.

Investigations of flight dynamics reveal that ragweed and most angiosperm pollen flies in the realm of very low Reynolds numbers where viscous forces will dominate inertial forces. In this situation pollen surface patterning will affect flight parameters to the degree that it changes the pollen grains effective density and radius. One way that the sculpture can change effective density is by changing the degree to which the pollen grains stick together and form clumps.

Recent studies of animal pollinated plants (FERGUSON 1984; FERGUSON & SKVARLA 1981, 1982; HEMSLEY & FERGUSON 1985; HESSE 1979 a, b, c, 1980 a, b, 1981 a, b, 1984 a, b; SKVARLA & al. 1978 a; WAHA 1984) have begun to elucidate the role of pollen surface sculpture in the control of pollenkitt distribution and thus pollen "stickiness", and the role of viscin threads and other thread-like structures in the regulation of pollen clumping. HESSE has also studied the differences in pollenkitt and pollen surface structure in comparisons of related wind- and animal-pollinated taxa (HESSE 1979 a, b, c, 1980 a). These studies have shown that surface sculpture does interact with pollenkitt distribution to control pollen adhesion.

More direct effects of surface sculpture on flight dynamics are less easy to evaluate. Theoretically, the importance of surface roughness should decrease with the Reynolds number and at those Re that govern pollen flight dynamics, be of minimal importance (HUTCHINSON 1967: 270; VOGEL 1981: 83). This, and HUTCH-INSON (1967: 270) makes the point quite clear, assumes that the variations in surface pattern do not affect "any essential change in shape". To this must be added the additional qualifications implicit in Hutchinson's equations, that for the effects of surface roughness to be negligible, it must not change either the effective density or radius. Calculations based on measurements of ragweed pollen indicate that the degree of surface roughness created by even the relatively short spines exceeds these limits; the spines increase both the effective density and radius. Our concepts of roughness must also be scaled; the type of pollen surface sculpture with aerodynamically insignificant roughness is most likely to be in the scabrate-psilate range. Larger sculptural elements ($>1\,\mu m$) are likely to produce a significant increase in radius for grains with diameters typical of anemophilous plants. (Pollen diameters of anemophilous species are typically 20 to $40\,\mu m$, Whitehead 1983; so sculptural elements larger than $1\,\mu m$ would increase the radius by 5 to 10%.)

Given the density of sporopollenin, an increase in surface sculpture must incorporate a large (c. 70 to 77%) volume of air to reduce the grain's effective density and thus enhance flight dynamics. Only the different types of pollen air bladders would seem to be an efficient way to do this.

Engineering studies of aerosol filtration suggest that surface features might affect pollen retention on the stigma. These studies show that rough particles are more likely to be relofted after deposition because they have less surface area in contact with the filter. The results of wind tunnel experiments on relofting of long and short spined *Compositae* pollen do not suggest that this is a major consideration in anemophily, perhaps because even highly ornamented pollen is relatively smooth and regular when compared to the particles considered rough in aerosol filtration.

Niklas's work (Niklas 1981 a, b, 1982, 1983, 1984, 1985 a, b, c, 1987; Niklas & Buchmann 1987; Niklas & al. 1986; Niklas & Norstog 1984; Niklas & Paw U. 1982, 1983) has shown that the morphology of the plant parts around the pollen receptive surfaces plays a critical role in determining patterns of pollen deposition. In general, these structures break up and slow the air flow around stigmas and create conditions more favorable for inertial impaction of pollen. In nature, with these lower air speeds and thus thicker boundary layers around stigmas, loss of pollen by relofting from the stigma is even less likely than it was under the experimental conditions of this study.

The role of electrostatic forces in pollen adhesion, discussed by Crane (1986), is an area that should prove interesting in relation to surface sculpture. Do some surface features such as large spines, act as charge collectors?

Smooth pollen may be advantageous in wind-pollination primarily as it serves to decrease pollen clumping and secondarily as it allows reduction in effective radius without an overall thinning of the exine. A corollary is that a particular surface sculpture may not correlate directly with transmission by a particular pollination vector on the basis of aerodynamic efficiency or some lock-and-key fit. Any sculpture pattern that produces the same effective density, radius, and stickiness (through regulation of pollenkitt deposition), will serve in place of another.

A secondary factor leading to loss of ornamentation in anemophilous plants may be the energetic cost of sporopollenin. Sporopollenin, a lipid based substance (Brooks & Shaw 1968), is energetically expensive (Penning de Vries & al. 1974) when compared to other wall materials. Smooth pollen surfaces may result, in part, from selection for metabolically "cheaper" pollen walls.

This work was funded by N. S. F. award R 11-8503107 to M. R. BOLICK. I thank the Departments of Botany and Zoology at Duke University for their hospitality. Special thanks go to the BLIMP Seminar, D. A. LIVINGSTONE, S. VOGEL, and S. A. WAINWRIGHT.

References

BODMER, H., 1922: Über den Windpollen. – Natur u. Tech. Zürich **3**: 66.

BROOKS, J., SHAW, G., 1968: The post-tetrad ontogeny of the pollen wall and the chemical structure of the sporopollenin of *Lilium henryi*. – Grana Palynol. **8**: 227–234.

CAMAZINE, S., NIKLAS, K. J., 1984: Aerobiology of *Symplocarpus foetidus:* interactions between the spathe and spadix. – Amer. J. Bot. **71**: 843–850.

COCKE, E. C., 1937: Calculating pollen concentration of the air. – J. Allergy **8**: 601–606.

CRANE, P. R., 1986: Form and function in wind dispersed pollen. – In BLACKMORE, S., FERGUSON, I. K., (Eds.): Pollen and spores: form and function. – Linn. Soc. Symp. Ser. **12**: 179–202.

DURHAM, O. C., 1946: The volumetric incidence of atmospheric allergens. 3. Rate of fall of pollen grains in still air. – J. Allergy **17**: 70–78.

DYAKOWSKA, J., 1937: Researches on the rapidity of the falling down of pollen of some trees. – Bull. Int. Acad. Cracovie (Acad. Po. Sci.), ser. B, Sci. Nat. **1**: 155–168.

FAEGRI, K., IVERSEN, J., 1975: Textbook of pollen analysis. 3rd edn. – New York: Hafner.

– VAN DER PIJL, L., 1979: The principles of pollination ecology. 3rd edn. – Oxford: Pergamon.

FERGUSON, I. K., 1984: Pollen morphology and biosystematics of the subfamily *Papilionoideae* (*Leguminosae*). – In GRANT, W. F., (Ed.): Plant biosystematics, pp. 377–394. – Toronto: Academic Press.

– SKVARLA, J. J., 1981: The pollen morphology of the subfamily *Papilionoideae* (*Leguminosae*). – In POLHILL, R. M., RAVEN, P. H., (Eds.): Advances in legume systematics, pp. 859–896. – Kew: Royal Botanic Gardens.

– – 1982: Pollen morphology in relation to pollinators in *Papilionoideae* (*Leguminosae*). – Bot. J. Linn. Soc. **84**: 183–193.

FRANKEL, R., GALUN, E., 1977: Pollination mechanisms, reproduction and plant breeding. – New York, Berlin, Heidelberg: Springer.

GREGORY, P. H., 1973: The microbiology of the atmosphere. 2nd edn. – Aylesbury, England: Leonard Hill.

HARRINGTON, J. B., METZGER, K., 1963: Ragweed pollen density. – Amer. J. Bot. **50**: 532–539.

HEMSLEY, A. J., FERGUSON, I. K., 1985: Pollen morphology of the genus *Erythrina* (*Leguminosae: Papilionoideae*) in relation to floral structure and pollinators. – Ann. Missouri Bot. Gard. **72**: 570–590.

HESSE, M., 1979 a: Ultrastruktur und Verteilung des Pollenkitts in der insekten- und windblütigen Gattung *Acer* (*Aceraceae*). – Pl. Syst. Evol. **131**: 277–289.

– 1979 b: Entwicklungsgeschichte und Ultrastruktur von Pollenkitt und Exine bei nahe verwandten entomo- und anemophilen Angiospermen: *Polygonaceae*. – Flora **168**: 558–577.

– 1979 c: Entstehung und Auswirkungen der unterschiedlichen Pollenklebrigkeit von *Sanguisorba officinalis* und *S. minor*. – Pollen & Spores **21**: 399–413.

– 1980 a: Ultrastruktur und Entwicklungsgeschichte des Pollenkitts von *Euphorbia cyparissias, E. palustris* und *Mercuralis perennis* (*Euphorbiaceae*). – Pl. Syst. Evol. **135**: 253–263.

– 1980 b: Zur Frage der Anheftung des Pollens an blütenbesuchende Insekten mittels Pollenkitt und Viscinfäden. – Pl. Syst. Evol. **133**: 135–148.

– 1981 a: The fine structure of the exine in relation to the stickiness of angiosperm pollen. – Rev. Palaeobot. Palynol. **35**: 81–92.

HESSE, M., 1981 b: Pollenkitt and viscin threads: their role in cementing pollen grains. −
    Grana **20**: 145−152.
−   1984 a: Form and function of *Delonix* pollen surface. − Mikroskopie **41**: 70−72.
−   1984 b: An exine architecture model for viscin threads. − Grana **23**: 69−175.
HUTCHINSON, G. E., 1967: A treatise on limnology, 2. Introduction to lake biology and
    the phytoplankton. − New York: Wiley.
KNOLL, F., 1932: Über die Fernverbreitung des Blütenstaubes durch den Wind. − Forsch.
    Fortschr. **8**: 301−302.
LEE, S. T., 1978: A factor analysis study of the functional significance of angiosperm pollen.
    − Syst. Bot. **3**: 1−19.
LEWIS, W. H., 1977: Pollen exine morphology and its adaptive significance. − Sida **7**:
    95−102.
NIKLAS, K. J., 1981 a: Simulated wind pollination and airflow around ovules of some early
    seed plants. − Science **211**: 275−277.
−   1981 b: Airflow patterns around some early seed plant ovules and cupules: Implications
    concerning efficiency in wind pollination. − Amer. J. Bot. **68**: 635−650.
−   1982: Simulated and empiric wind pollination patterns of conifer ovulate cones. −
    Proc. Natl. Acad. Sci. U.S.A. **79**: 510−514.
−   1983: The influence of Paleozoic ovule and cupule morphologies on wind pollination.
    − Evolution **37**: 968−986.
−   1984: The motion of windborne pollen grains around conifer ovulate cones: implications
    on wind pollination. − Amer. J. Bot. **71**: 356−374.
−   1985 a: Wind pollination of *Taxus cuspidata*. − Amer. J. Bot. **72**: 1−13.
−   1985 b: The aerodynamics of wind pollination. − Bot. Rev. **51**: 328−386.
−   1985 c: Wind pollination−a study in controlled chaos. − Amer. Sci. **73**: 462−470.
−   1987: Pollen capture and wind-induced movement of compact and diffuse grass panicles:
    implications for pollination efficiency. − Amer. J. Bot. **74**: 74−89.
−   BUCHMANN, S. L., 1985: Aerodynamics of wind pollination in *Simmondsia chinensis*
    (LINK) SCHNEIDER. − Amer. J. Bot. **72**: 530−539.
−   − 1987: The aerodynamics of pollen capture in two sympatric *Ephedra* species. −
    Evolution **41**: 104−123.
−   − KERCHNER, V., 1986: Aerodynamics of *Ephedra trifurcata:* 1. Pollen grain velocity
    fields around stems bearing ovules. − Amer. J. Bot. **73**: 966−979.
−   KERCHNER, V., 1986: Aerodynamics of *Ephedra trifurcata*. 2. Computer modelling of
    pollination efficiencies. − J. Mathemat. Biol. **24**: 1−24.
−   NORSTOG, K., 1984: Aerodynamics and pollen grain depositional patterns on cycad
    megastrobili: implications on the reproduction of three cycad genera (*Cycas, Dioon,*
    and *Zamia*). − Bot. Gaz. **145**: 92−104.
−   PAW U., K. T., 1982: Pollination and airflow patterns around conifer ovulate cones.
    − Science **217**: 442−444.
−   − 1983: Conifer ovulate cone morphology: implications on pollen impaction patterns.
    − Amer. J. Bot. **70**: 568−577.
PAYNE, W. W., SKVARLA, J. J., 1970: Electron microscope study of *Ambrosia* pollen (*Compositae: Ambrosieae*). − Grana **10**: 89−100.
PENNING DE VRIES, F. W. T., BRUNSTING, A. H. M., VAN LAAR, H. H., 1974: Products,
    requirements and efficiency of biosynthesis: a quantitative approach. − J. Theor. Biol.
    45: 339−377.
PRAGLOWSKI, J., GRAFSTRÖM, E., 1980: The pollen morphology of the tribe *Calenduleae*
    with reference to taxonomy. − Bot. Not. **133**: 177−188.
PROCTOR, M., YEO, P., 1972: The pollination of flowers. − New York: Taplinger Publishing
    Co.

RAYNOR, G. S., OGDEN, E. C., HAYES, J. V., 1970: Dispersion and deposition of ragweed pollen from experimental sources. − J. Appl. Meteorol. **9**: 885−895.

ROBBINS, R. R., DICKINSON, D. B., RHODES, A. M., 1979: Morphometric analysis of pollen from four species of *Ambrosia* (*Compositae*). − Amer. J. Bot. **66**: 538−545.

SCHEPPEGRELL, W., 1917: Hay-fever and hay-fever pollens. − Arch. Internal Med. **19**: 959−980.

SKVARLA, J. J., LARSON, D. A., 1965: An electron microscopic study of pollen morphology in the *Compositae* with special reference to the *Ambrosiinae*. − Grana Palynol. **6**: 210−269.

− RAVEN, P. H., CHISSOE, W. F., SHARP, M., 1978a: an ultrastructural study of viscin threads in *Onagraceae* pollen. − Pollen & Spores **20**: 5−143.

− TURNER, B. L., PATEL, V. C., TOMB, A. S., 1978b: Pollen morphology in the *Compositae* and in morphologically related families. − In HEYWOOD, V. H., HARBORNE, J. B., TURNER, B. L., (Eds.): The biology and chemistry of the *Compositae*, pp. 141−248. − New York: Academic Press.

SOKAL, R. R., ROHLF, F. J., 1981: Biometry. 2nd edn. − San Francisco: W. H. Freeman.

SOUTHWORTH, D., 1988: Isolation of exines from gymnosperm pollen. − Amer. J. Bot. **75**: 15−21.

VOGEL, S., 1981: Life in moving fluids. − Boston: Willard Grant.

WAHA, M., 1984: Zur Ultrastruktur und Funktion pollenverbindender Fäden bei *Ericaceae* und anderen Angiospermenfamilien. − Pl. Syst. Evol. **147**: 189−203.

WHITEHEAD, D. R., 1969: Wind pollination and the angiosperms: evolutionary and environmental considerations. − evolution **23**: 28−35.

− 1983: Wind pollination: some ecological and evolutionary perspectives. − In REAL, L., (Ed.): Pollination biology, pp. 97−108. − New York: Academic Press.

WODEHOUSE, R. P., 1935: Pollen grains. − New York: McGraw-Hill.

Address of the author: MARGARET R. BOLICK, W-532 Nebraska Hall, University of Nebraska State Museum, University of Nebraska, Lincoln, NE 68588-0514, U.S.A.

Pl. Syst. Evol. [Suppl. 5], 53−69 (1990)

Harmomegathic characters of *Pteridophyta* spores and *Spermatophyta* pollen

E. Pacini

Received February 3, 1988

Key words: *Pteridophyta, Spermatophyta.* − Spore, pollen, harmomegathy.

Abstract: Spore and pollen grain volume decrease after meiotic cleavage and starts to increase during exine and intine formation; with dehydration volume decreases again. In the atmosphere pollen volume accomodates according to relative humidity. The mechanisms leading to and regulating dehydration of the sporangia and anthers, the stationary phase during dispersal, and hydration are species specific. They vary between taxonomic groups and are influenced by the environment. Pollen adhesion occurs only in angiosperms and is a prerequisite for pollen hydration; the opposite is true for gymnosperms. The environment of the landing site determines physical recognition which enables proper hydration to take place. Another instance of physical recognition is found in fern spores before germination, and involves temperature and light quality. Chemical recognition appears only in *Spermatophyta*. The water for rehydration is provided in different was in pteridophytes, gymnosperms and angiosperms, and the molarity of the hydrating solutions varies. In vitro hydration tests have demonstrated that *Pteridophyta* spores achieve maximum size in water, whereas the optimum media for gymnosperms and angiosperms have sucrose concentrations of 5−50 g/l. The volume of *Spermatophyta* pollen grains decreases during germination but that of *Pteridophyta* spores increases. The increase percentages in volume from the dry to the hydrated state vary between species and depend on the hydration site and the spore/grain itself. The internal factors are: hydration of the intine, selective permeability of the plasma membrane, hydration state of the cytoplasm, and accomodating power of the walls. Different devices regulate these volumetric changes in angiosperm pollen grains, e.g., furrows and apertures. Harmomegathic devices are less variable in gymnosperm that in angiosperm pollen, and even less variable in the pteridophytes.

Harmomegathy is a word coined by WODEHOUSE (1935) to indicate the changes in pollen volume and shape during dehydration and hydration. Even the spores of land plants dehydrate before being dispersed and rehydrate before germination. Dehydration and rehydration are complex phenomena because they depend on pollen/spore structure, the mother plant and the environment where the pollen/spore lands (PACINI 1986).

Pollen grains and spores are subject to mechanical stress during dehydration and rehydration, when their walls and plasma membranes permit the passage of water (HESLOP-HARRISON 1979 a, b). The exine is more involved than the intine in

relative physical adaptations because of its elasticity. Exine elasticity is also demonstrated by the fact that it is the first to be formed and has to bear the increase
in microspore volume until intine formation. In *Selaginella kraussiana* the total
volume of the exine of the megaspore increases by a factor of 1.5×10^4 from the
tetrad to the mature stage (BUCHEN & SIEVERS 1981). Despite a large increase in
the volume of the *Lilium* spore the exine thickness remains constant (WILLEMSE &
REZNICKOVA 1980).

Sporopollenin occurs in some in green algae and in the spore and pollen grain
walls of all land plants (PACINI & al. 1985) with the exception of a few angiosperms
(KRESS 1986). Evidence that sporopollenin is a determinant for dehydration and
rehydration comes from the fact that the exine is extremely reduced, discontinuous
or even absent in marine monocot pollen (DUCKER & al. 1978). The extreme
adaptation of these plants to submarine life and the lack of dehydration lead to
the disappearance of the sporopollenin around the pollen grains.

Water content ranges widely $(10-50\%)$ in angiosperm pollen shed. This mostly
has been measured in anemophilous species and varies widely from author to author
and even in with the same species, probably due to different relative humidity (RH)
at the time of shedding (PACINI & al. 1988). Water content in pollen is often measured
by gravimetric analysis. Recently DUMAS & GAUDE (1983) proposed the NMR
method and sought for a correlation between water content and pollen viability.

The walls, plasma membrane and protoplast of spores and pollen grains are
structured so as to bear, within certain limits, the stresses created by dehydration
and hydration. There must also be some other mechanisms to reduce harmomegathic changes during dispersal. Quite recently, BLACKMORE & BARNES (1986) and
THANIKAIMONI (1986) have reviewed the harmomegathic mechanisms in pollen
grains, mostly furrows and apertures. These devices have received less study in fern
spores but the laesura, a "proximal mark with the exine normally folded outward
into a tent-like crest" (THANIKAIMONI 1978) has the function of an aperture.

Dehydration-hydration naturally occurs only once for a pollen grain but according to HESLOP-HARRISON (1979 a, b) "the cycle may be repeated many times
in experiments without impairing the viability of the pollen".

Pollen and spores are subjected to changes in water content during their journey
through the atmosphere and the exine prevents extreme moisture losses or gains.
The RH at shedding and during pollen transport is responsible for these changes
and for pollen viability; each species reacts in a different way and has its own
pattern.

The spores and pollen grains land on a surface at the end of their journey and
a process of physical recognition begins. The molarity of the rehydration liquid
and water availability determines the success or the failure of regocnition. The right
places for rehydration are different for each type of reproductive structure. Fern
spores rehydrate in the water of the soil; gymnosperm pollen in the micropylar
drop or the stigmatic micropyle, and angiosperm pollen on the stigma. Quite
recently, NICKLAS (1985) has demonstrated that in gymnosperms a preliminary
physical recognition between the stereostructure of pollen and that of female cones
occurs before the pollen lands.

The study of harmomegathy can be approached by cytophysiological methods,
i.e., by observing the changes in volume, shape, and pollen viability at different
RHs. For certain aspects such as the morphology of the plasma membrane with

the cytoplasm in a dehydrated or hydrated state, cytological methods as, e.g., freeze etching are currently used (PLATT-ALOIA & al. 1986).

Materials and methods

The cytological routine techniques used in cytoembryological research can not be applied to study harmomegathy. Only techniques such as SEM and pollen grain viability may be used. Therefore, new techniques have been devised to observe pollen volume and changes in shape.

Plant material. Pollen grains and spores for the experiments were collected from plants cultivated in the Botanical Gardens of Siena University. Pollen sacs and stamens were vigorously shaken to release the pollen. The spores were collected with a needle to induce the opening of the sporangia. The spores and pollen grains were collected a few seconds before each experiment. Only spores and pollen grains of high viability (> 80%) were used in the experiments; viability was determined by the fluorochromatic reaction test procedure (HESLOP-HARRISON & al. 1984). For *Pteridophyta* spores and gymnosperm pollen prehydration was necessary to induce the fluorochromatic reaction.

Pteridophyta spores were collected from sporangia of *Equisetum telmateia* EHRH., *Phyllitis scolopendrium* (L.) NEWMAN; *Platycerium alcicorne* (WILLEM.) TARDIEAU, *Polystichum lonchitis* (L.) ROTH. and *Psilotum nudum* (L.) P. BEAV. Gymnosperm pollen was collected from the pollen sacs of *Ephedra americana* HUM. & BOMPL., *Podocarpus macrophylla* D. DON. and *Taxus baccata* L. Angiosperm pollen was collected from anthers of *Capparis spinosa* L., *Centranthus ruber* (L.) DC., *Diplotaxis erucoides* (L.) DC., *Helleborus foetidus* L., *H. viridis* L., *Hypericum calycinum* L., *Juglans regia* L., *Ornithogalum umbellatum* L., *Parietaria judaica* L., and *Primula vulgaris* HUDSON.

In vitro pollen hydration. The changes in volume and shape of pollen grains and spores can be simulated in vitro using solutions containing $5-200$ g/l of sucrose (PACINI 1986). Newly shed pollen grains and spores were dusted on slides and immersed in a sucrose solution and covered with a cover slide. Two more slides were prepared with water and immersion oil respectively. The latter slide gives the size of dry pollen.

Each slide was observed after the same interval of time from preparation to avoid errors in case equilibrium had not yet been reached between pollen/spore and medium.

Juglans regia pollen grains were kept at 80 °C for 24 h to make them unviable (as assessed by the fluorochromatic reaction). Their behaviour was compared to living grains, those cytoplasm plays a role during hydration.

The dimensions: diameter or longitudinal and equatorial axes were recorded using a lanometer projecting microscope. High magnification was used to check protoplast plasmolysis. When pollen shape resembled a geometric figure the volume was calculated by the formulas $^4/_3 \pi r^3$ for spheres and $^4/_3 \pi a^2 b$ for ellipsoids where a is the equatorial axis/2 and b is the longitudinal axis/2. The sizes of 100 grains/spores were scored for each slide. A graph was prepared for each species with sugar concentration as abscissa and volume or size as ordinate.

It was assumed that the maximum size reached during hydration in vitro was similar to that reached before germination on a compatible stigma or micropyle.

Effects of different relative humidities on pollen and spore size and viability. Parallelepipeds cut from a bar used to prepare glass ultramicrotome knives were placed in petri dishes. Saturated solutions of different salts giving appropriate RH (SHIVANNA & JOHRI 1985) were added. Newly shed pollen/spore grains were dusted on the upper surface of the glass parallelepipeds and immediately closed with petri dish covers. The relative humidity of the flower or sporangia environments was recorded before pollen/spore shedding and pollen/spore volume was measured at the beginning of each experiment. More than one petri dish was prepared for each relative humidity and grains from each one were used to check

Table 1. Pollen volume after intraspecific and interspecific pollination. × Germination on an interspecific stigma

Pollen	Stigma		
	Centranthus ruber	*Diplotaxis erucoides*	*Hypericum calycinum*
Centranthus ruber (*Valerianaceae*)	107.715	164.636	−
Diplotaxis erucoides (*Cruciferae*)	8.728	11.993	−
Hypericum calycinum (*Hypericaceae*)	5.818 ×	6.714	6.538
Parietaria judaica (*Urticaceae*)	2.144 ×	−	3.769

pollen/spore viability by the fluorochromatic reaction and pollen volume after different intervals of time from the beginning of the experiment.

Pollen volume after self and cross pollination experiments. Flower buds were brought to the laboratory the day before anthesis and emasculated. Self compatible pollinations and cross pollinations were performed. Pollen size was detected by plunging the stigma into immersion oil after 5′, 15′, 30′; 1 h, 2 h, 3 h. 30 − 50 pollen grains were scored and measured for each time interval. Only pollen grains attached to the stigmatic papillae by a colpus were measured. The maximum volume reached by pollen grains on their own and foreign stigma is reported in Table 1.

Observations

Spores and pollen grains decrease in volume after meiotic cleavage, they increase in size during exine and intine formation, decrease again during sporangia and anther dehydration (Fig. 1). Pollen volume is not constant during pollen dispersal but fluctuates according to RH (Fig. 1). The right landing sites for fern spores, gymnosperm and angiosperm pollen grains are different but physical recognition always occurs.

Harmomegathy in *Pteridophyta* spores. The maximum size of *Equisetum telmateia* spores (Fig. 2) like those of the other *Pteridophyta* (see Materials and methods) is attained in water. *Equisetum* spore volume decreases with increasing sucrose concentrations (Fig. 2) though not in a linear way. The perine breaks only when spores of this type are plunged into water. The volume of *Equisetum telmateia* spores kept at room temperature and 45% RH decreases seven hours after opening of the sporangia, even if the spores are still viable.

Harmomegathy in gymnosperms. The sacci of *Podocarpus macrophylla* pollen diverge during hydration as the diameter of the grain containing the protoplast increases. The sacci do not increase in volume because they are waterproof; water penetrates only after washing with lipid solvents. Plasmolysis starts to occur at 100 g/l sucrose concentration (Fig. 3). Dry *Taxus baccata* pollen has a shrivelled appearance with thick intine and thin exine. During hydration, but before the exine

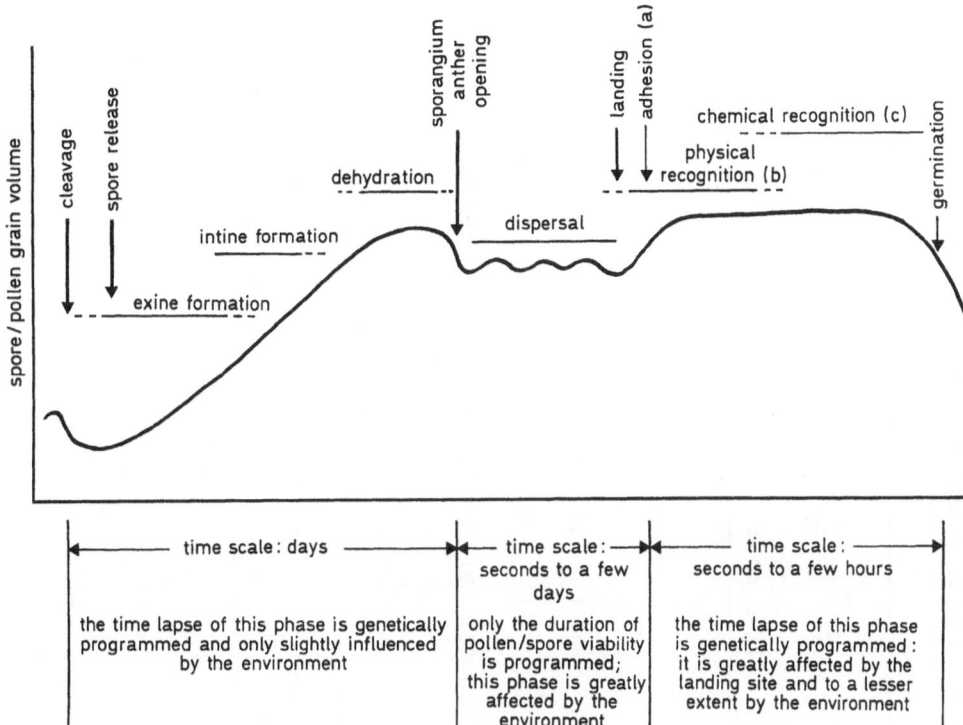

Fig. 1. Scheme summarizing fern spore and pollen grain volume changes during development and after dispersal. Fern spores and pollen grains decrease in volume after release from the tetrads; they increase during exine and intine formation; they decrease during sporangia and anther dehydration. Their changes in shape, due to dehydration – hydration, depend on spore/pollen geometry and sporoderm structure. During dispersal the volume accommodates according to environmental relative humidity (RH). Landing sites regulate the hydration of spore/pollen and determinate physical recognition. Pollen grains decrease in size during germination but fern spores do not. (a) Pollen adhesion occurs only in angiosperm pollen; (b) secondary physical recognition occurs in fern spores before germination; (c) chemical recognition occurs only in *Spermatophyta*.

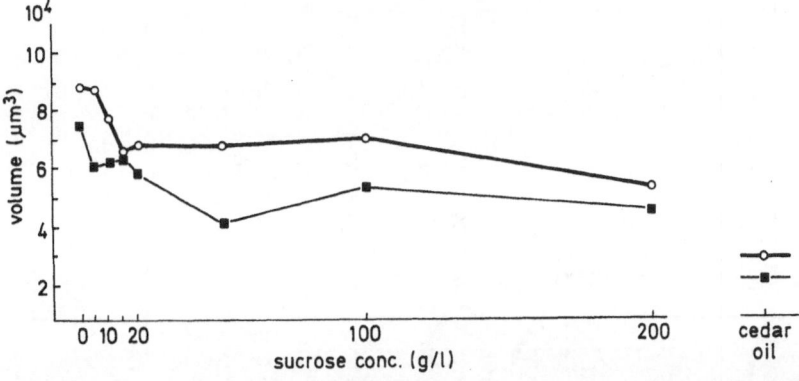

Fig. 2. Changes in volume of *Equisetum telmateia* spores in increasing sucrose concentrations. The experiment was performed twice, with newly released spores (○) and 7 h (■) later with spores kept at room temperature

Table 2. Hydration state percentages of *Taxus baccata* pollen grains in increasing sucrose concentrations and at different time lapses (1', 15', 12 h)

Dry grain — Hydrated grains

Sucrose concentration (g/l)	1'	15'	12 h
0	5.5 / 94.5	79.5 / 21.5	12.3 / 37.4 / 49.8
5	100	68.7 / 31.3	68.4 / 31.6
10	100	100	55.7 / 22.3 / 22.0
15	8.3 / 91.7	100	68.2 / 31.8
20	15.1 / 84.9	3.2 / 96.8	100
50	81.1 / 18.9	13.4 / 74.2 / 12.4	78.6 / 21.4
100	11.3 / 7.6 / 60.0 / 21.1	12.4 / 62.2 / 24.3	11.9 / 64.5 / 24.6
200	86.9 / 13.1	75.7 / 24.3	34.2 / 63.2 / 2.6

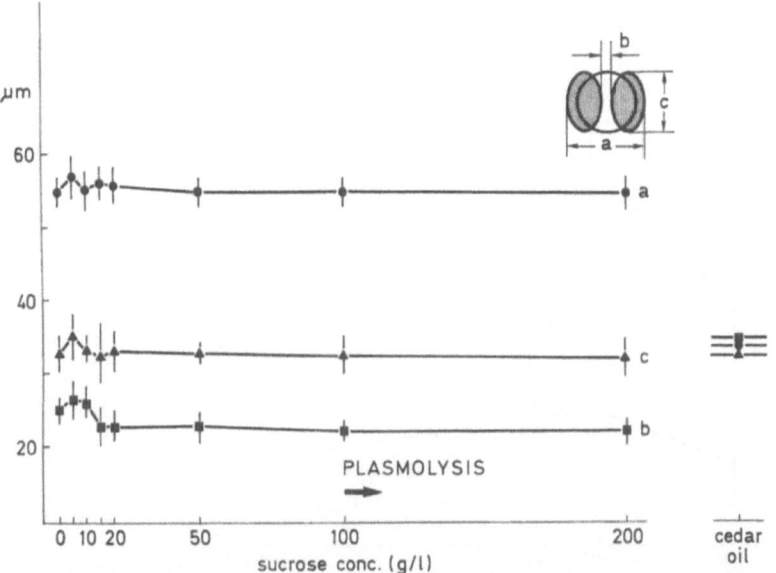

Fig. 3. Changes in size of *Podocarpus macrophylla* pollen grains in increasing sucrose concentrations. Plasmolysis starts to occur at 100 g/l sucrose concentration

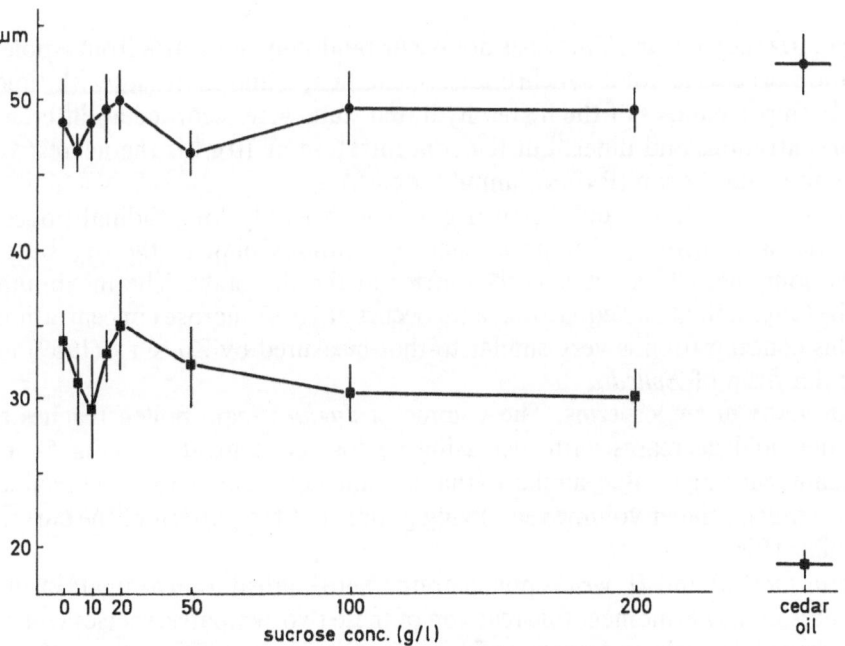

Fig. 4. Changes in longitudinal (—●—) and equatorial (—■—) axes of *Ephedra americana* pollen grains at different sucrose concentrations; maximum size at 20 g/l

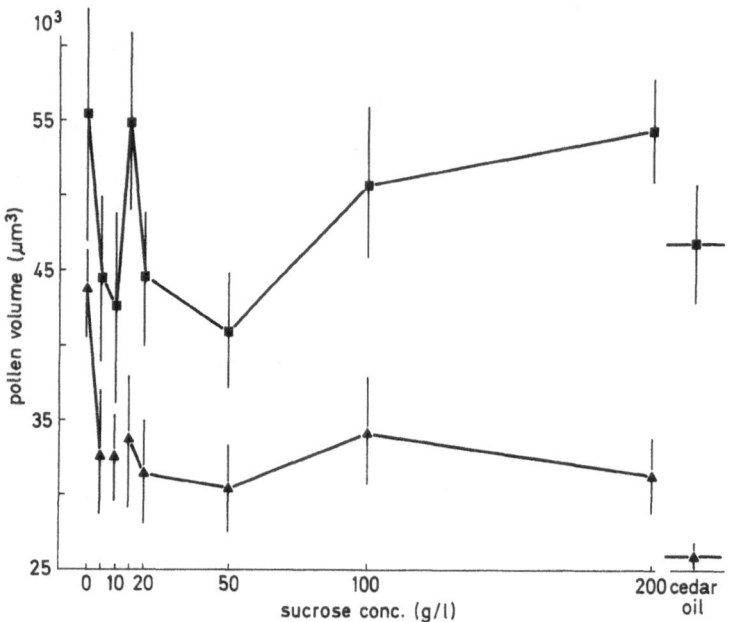

Fig. 5. Changes in volume of living (—■—) and unviable (—▲—) *Juglans regia* pollen in increasing sucrose concentrations

ruptures, polarization occurs. This does not occur randomly but starts from a pole. The hydration process is not a synchronous phenomenon and increases with time (Table 2). High percentages of the higher hydrated state were recorded at different sucrose concentrations and times, but the concentration of 10 g/l is the nearest to that of the pollination drop (PACINI, unpubl. data).

Ephedra americana has ovoidal pollen grains with empty longitudinal ridges. The equatorial axis is always longer in sucrose solutions than in the dry state, whereas the longitudinal axis is slightly shorter in the dry state. The maximum length of the longitudinal and equatorial axes occurs at 20 g/l sucrose concentration (Fig. 4). This concentration is very similar to that measured by ZIEGLER (1959) in the micropylar drop of *Ephedra*.

Harmomegathy in angiosperms. The volume of *Juglans regia* pollen reaches a peak in water, and decreases with increasing sucrose concentrations (Fig. 5). It increases again reaching a value similar to that obtained in water at 15 g/l. Unviable grains have a much smaller volume than living grains, but the pattern of the curves is quite similar (Fig. 5).

Helleborus foetidus and *H. viridis* pollen grains were studied to investigate eventual differences in the harmomegathic reaction of these two sympatric species which bloom contemporaneously and compete for pollinating insects. The pattern of the curves of these species is quite similar with three peaks (Fig. 6), but the maximum size is reached at 15 g/l by *H. foetidus* and at 50 g/l by *H. viridis*.

The volume of *Ornithogalum umbellatum* pollen grains accomodates according to RH and time of exposure (Fig. 7). After 30′ of 98% RH it increases in volume but to a lesser extent at 15, 51, and 79% RH (Fig. 7). After 1 h 15′ the volume of pollen grains kept at 15% RH is the same as for 30′ but decreased for the other

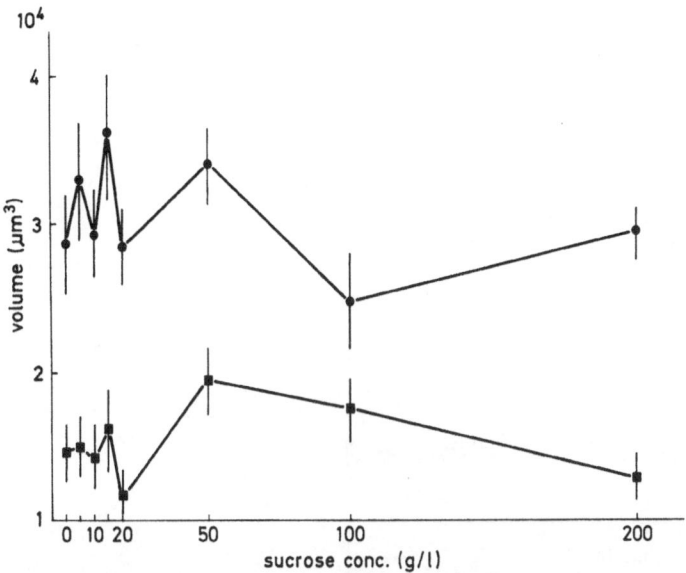

Fig. 6. Changes in volume of *Helleborus foetidus* (—■—) and *H. viridis* (—●—) pollen grains in increasing sucrose concentrations

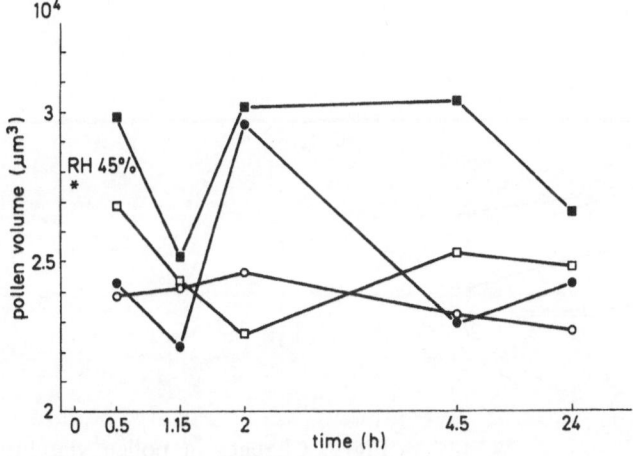

Fig. 7. Changes in volume in *Ornithogalum umbellatum* pollen grains kept at different RHs (—○— 15, —●— 51, —■— 79, and —□— 98%) for different time lapses. ★ Pollen volume recorded before the start of the experiment

three RH. As time proceeds the volume of pollen grains kept at 98%RH and 51% RH increases again (Fig. 7).

The behaviour of the pollen grains produced by the pin and thrum flowers of *Primula vulgaris* is vastly different as is their response to the different RHs (Fig. 8). After 1 h at 4 different RHs the volume of thrum pollen grains was less than at the start of the experiment. Pin pollen grains kept at 98% RH increased in volume. After 2 h thrum pollen grains had increased in volume with the maximum at 79% RH.

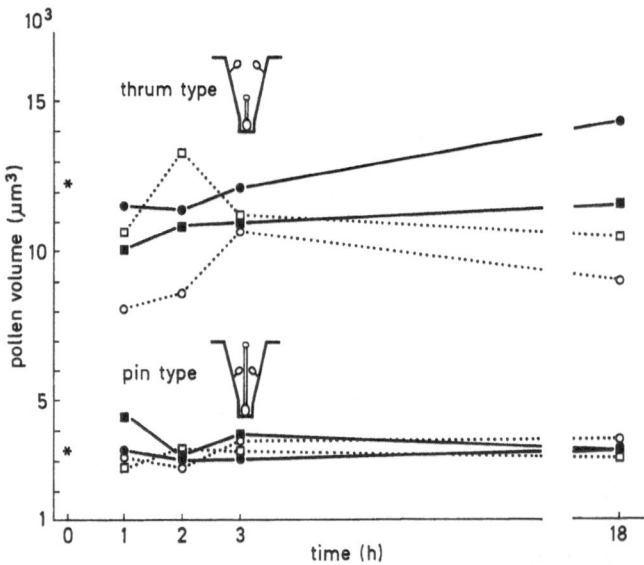

Fig. 8. Changes in volume of *Primula vulgaris* thrum and pin pollen grains at different RHs (·· ○ ·· 15, —●— 52, ··□·· 79, and —■— 98%) for different time lapses. ★ Pollen volumes of the two forms before the start of the experiment

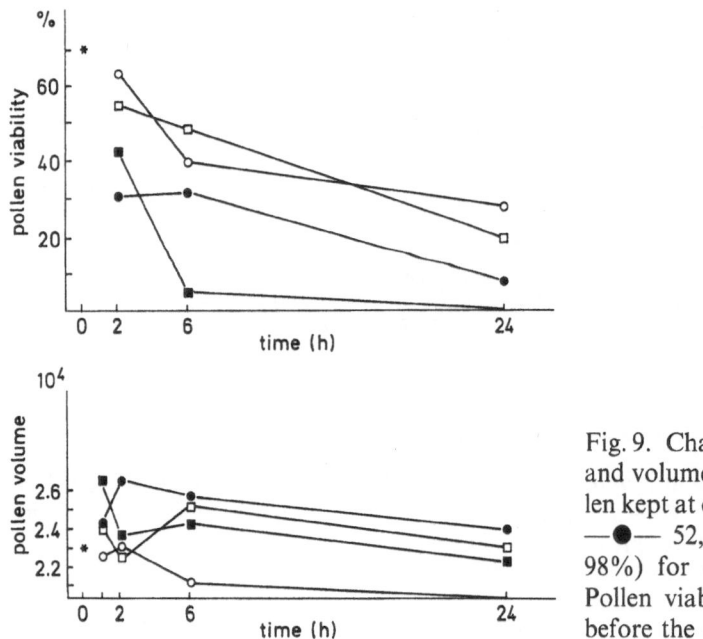

Fig. 9. Changes in pollen viability and volume of *Capparis spinosa* pollen kept at different RHs (—○— 15, —●— 52, —□— 79, and —■— 98%) for different time lapses. ★ Pollen viability and pollen volume before the start of the experiment

Relative humidity influences the viability of airborne pollen and pollen exposed in the anther. Pollen grain volume and viability were determined in the same samples of *Capparis spinosa* pollen kept at four RHs for different periods of time (Fig. 9). The viability was already reduced 2 h after the start of the experiment at the four different RHs. The best viability was recorded for pollen grains kept at 15% RH and the lowest at 50% RH (Fig. 9). After 24 h the pollen grains kept at 98% RH were unviable and a small percentage of those kept at 15, 50, and 79% RH had survived (maximum survival at 15% RH). The volume of *C. spinosa* pollen kept at the tested RHs for 2 h increased at 51, 79, and 98% RH but slightly decreased at 15% RH. After 4 h the volume of the pollen grains kept at 79 and 98% RH had decreased but that kept at 15 and 50% RH had increased. New increases occurred after 6 h and after that the volume decreased. The pollen grains kept at 15% RH had the highest viability and their volume was less than at the start of the experiment.

The increase in volume after intraspecific and interspecific pollination varied from species to species (Table 1). *Centranthus ruber* pollen absorbed more water from the *Diplotaxis* stigma than from its own; the opposite occurs in *Diplotaxis*. There is a low percentage of germination of *Hypericum calycinum* and *Parietaria judaica* pollen on the *Centranthus ruber* stigma. In these cases the volume reached on the foreign stigma was quite different (either greater or less) from that reached on its own (Table. 1).

Discussion

The peculiar wall specialization of fern spores and pollen grains protects them against damage during development, dehydration, the stationary phase, dispersal and rehydration. As soon as pollen grains and spores land on a surface, physical recognition between them and the surrounding liquid media occurs. If they land and hydrate in the correct way, i.e., if the molarity of the available liquid is correct, further physical recognition occurs for fern spores and chemical recognition for *Spermatophyta* (Fig. 1). Water is adequate for fern spore hydration. The optimum molarity for *Spermatophyta* pollen hydration varies. This probably depends on the fact that the landing sites are provided with liquids of different molarity and accessibility. Fern spores hydrate in environmental water and the molarity of the gymnosperm micropylar drop varies from species to species (SINGH 1978). The receptive surface of the stigma can be very different in angiosperms. Some are completely dry (without any exudate) while others have a lipidic or even liquid exudate or both (HESLOP-HARRISON & SHIVANNA 1977). The passage of water into the pollen grain is regulated both by the cuticle of the stigma papilla and the pollen grain apertures (HESLOP-HARRISON 1979 b).

The structures of fern spores and pollen grains involved in physical recognition are the exine, intine, plasma membrane and the partially dehydrated cytoplasm. Water passively crosses the exine without any changes in its structure, but as soon as it reaches the intine it is absorbed, provoking a first increase in volume, especially in the area where the intine is thicker, i.e., the aperture region. The plasma membrane acts as a selective barrier determining the absorption of water according to a program typical of each taxonomic group.

Plasma membrane structure in different states of hydration has only been investigated in angiosperms. PLATT-ALOIA & al. (1986) did not find any ultrastructural differences between dry and hydrated pollen grains of *Phoenix*, *Collomia* and *Zea*. The question of ultrastructural changes in the plasma membrane during dehydration and hydration is still much debated (HESLOP-HARRISON 1979 a, b; PLATT-ALOIA & al. 1986, KERHOAS & al. 1987), and needs separation from other morphological and physiological modifications.

The uptake of water by the intine and cytoplasm determines a second increase in volume of the spores and pollen grains up to an equilibrium determined by the elasticity of the exine and the initial water content of the grain. In both, gymnosperms and angiosperms, with an intine thicker than the exine, the maximum size of the protoplast and the maximum thickness of the intine are reached at different molarities. In water, there is more increase in intine thicknes, at higher sucrose concentrations there is more increase of cytoplasm. Water passively crosses the intine but when it reaches the plasma membrane the maximum absorption of water occurs at a well defined molarity. These differences in intine and protoplast water absorption explain the occurrence of more than one peak during the "in vitro" hydration of *Juglans regia* and *Helleborus* pollen grains. These species have a well developed oncus responsible for intine expansion which causes the increase in size of the pollen. The same phenomenon has also been detected in *Eucalyptus* spp. and hazel pollen (Heslop-Harrison & Heslop-Harrison 1985 a, Heslop-Harrison, Y. & al. 1986). When the intine is thinner or has the same thickness as the exine, this phenomenon is negligible.

An additional and very elaborate layer ("perine" or "perispore" according to Bower 1935) is superimposed on the fern exine. It consists of sporopollenin and is produced by the tapetum cytoplasm (Pettitt 1979). The loss of the perine by fern spores in water but not in sucrose solutions could mean that this layer helps the grain to keep its shape during hydration and be released only in the right hydration medium. *Equisetum telmateia* spores have no perine but their maximum hydration occurs in water.

Further evidence that water is the right medium for fern spore hydration and germination is provided by the fact that this is the medium commonly used to produce prothalli in vitro (Taylor & Luebke 1986).

Onoclea sensibilis spores have a very thick intine almost impermeable to solutes when dormant (Miller 1980). According to Miller the intine plays a major role in determining the passage of material into and out of the spore. The layer becomes permeable to low molecular weight solutes only after pretreatment with water.

The life-span of pollen and spores kept at environmental temperature rarely exceeds a few days. The record for longevity is held by the water fern *Marsilea* megaspores, which can produce gametophytes even at the age of 130 years (Johnson 1985). *Marsilea* megaspores are dispersed in sporocarps and have a very elaborate sporoderm (Machlis & Rawitscher-Kunkel 1967) which prevents the rapid entry or exit of water and maintains viability for a long time.

The harmomegathic behaviour of gymnosperm pollen differs according to its external geometry. Winged pollen grains show a germinal furrow which closes in a dry environment but opens in high humidity (Pettitt 1985). The sacci diverge during in vitro hydration due to the entry of water into the protoplast. But the shape of this kind of pollen must be reasonably constant at pollination because it is responsible for its correct conveyance to the micropyle, as elegantly demonstrated by Nicklas (1985) in tunnel experiments. In temperate environments, where the high humidity of the night cleans the atmosphere of all types of pollen, only freshly shed pollen can reach the microplylar drop.

The exine of angiosperm pollen can usually bear the increase in volume due to hydration even in water. In contrast, pollen grains of some gymnosperms, such as *Taxus baccata* (this study) and *Juniperus communis* (Duhoux 1982) with a shrivelled appearance when dry, have a very thick intine which remains around the pollen protoplast after the exine has broken during hydration. This occurs after the micropylar drop has withdrawn inside the mycropylar chamber next to the nucellus, where the pollen grains complete their development in a moist, isolated environment; the protoplast is thus not damaged. The drying of the pollination drop may help the pollen which is ready to germinate, to stick to the nucellar surface (SINGH 1978). The loss of the exine during hydration in several gymnosperm species means that at least in this group there is a very peculiar exine-intine interface which permits the exine to fall away. The rupture of the exine starts from a very well organized structure similar to a pore (DUHOUX 1982).

Gymnosperm pollen cell components at shedding vary among taxa, and range from one to more than ten cells. In multicellular pollen grains, the cell walls are always continuous with the intine (SINGH 1978) and dilate homogeneously during hydration. The generative cell and sperm cells of angiosperms on the other hand normally float inside the vegetative cell cytoplasm with which they expand during hydration (HESLOP-HARRISON, J. & al. 1987). A generative cell tail sometimes wraps the vegetative nucleus and may reduce the stresses of the "male germ unit" sensu DUMAS & al. (1984) during the process of dehydration and hydration. The existence of a generative cell tail anchored to the intine in the pollen grain tetrads of some *Rhododendron* spp. (THEUNIS & al. 1985) seems to strengthen this hypothesis. Such anchorage is in fact broken with hydration, before pollen tube emission (KAUL & al. 1987).

As soon as a pollen grain meets a compatible stigma it starts to adhere and the adhesion increases with water uptake. Adhesion is very poor in some cases of selfincompatibility (STEAD & al. 1979, 1980).

FERRARI & al. (1985) recognize four independent pollen-stigma binding forces, i.e., van der Waal, enzymatic, agglutination, and gelatinization binding forces staggered in time. The position of the pore with respect to the site of adhesion and the water content of the pollen determines the fate of the pollen and the speed of hydration (BARNABÁS & FRIDVALSKY 1984, BARNABÁS 1985).

The type of stigma determines the hydration mode. HESLOP-HARRISON & HESLOP-HARRISON (1982), in a survey of stigmatic cuticles, found that in the "dry" type, the cutin of the stigmatic papillae is not continuous but composed of small rods or globuli, disposed in a polysaccharide matrix. This discontinuous structure allows the water to pass from the papilla cytoplasm to the grain. The discontinuous cuticle of maize silk becomes densely packed when the silk dries and so prevents water loss (WESTGATE & BOYER 1986). Vice versa, in the "wet" stigma type, the cuticle is generally broken (HESLOP-HARRISON & HESLOP-HARRISON 1982). In either case, a very low RH can determine water loss if the cuticle has no means to prevent it.

After pollination of the stigma, an increase in exudate has been recorded in several species (see references quoted by MARGINSON & al. 1985). In *Acacia* this stimulus is produced even by unviable but not by washed grains (MARGINSON & al. 1985). This means that at least in these species, the water for rehydration is secreted just after pollination. The staggering in time of stigma secretions has also been recorded in *Oenothera organensis* by HESLOP-HARRISON & HESLOP-HARRISON (1985 b).

Juglans regia pollen, unviable after exposure to 80 °C for 24 h, has half the volume of viable grains. Hydration is passive, and the volumes at different sucrose concentrations are always higher than the dry volume. In living pollen some values are higher and some lower, and reflect the different responses of pollen components to the various sucrose concentrations. *Eucalyptus* pollen exposed to 60 °C for 24 h still retains one-third of its germination capacity, but only a small proportion survives 24 h at 70 °C (HESLOP-HARRISON & HESLOP-HARRISON 1985 a). Even *Juglans regia* pollen has a high power of survival: LUZA & POLITO (1987) have shown that even 22 days old pollen, with a very low water content regained its ability to germinate if incubated in a high RH environment.

Helleborous foetidus and *H. viridis* have similar hydration curves with peaks at the same sucrose concentrations. As they live in the same habitat and bloom contemporaneously, interspecific pollination could be successful.

Pollen grains can absorb environmental humidity when RH is high or loose water when it is low (GILISSEN 1977), as in *Ornithogalum umbellatum*, a species with great accomodation power.

Pin and thrum *Primula vulgaris* pollen grains react in different ways to RH, but even in this species there is some accomodation power. The range of thrum type pollen volume is much wider than that of the pin type. The former attain maximum volume after 2 h exposure at 79% RH and the latter after 1 h at 98% RH. These findings confirm those of SHIVANNA & al. (1983) that the thrum pollen of *Primula* is more sensitive to high RH.

The effect of different RHs on *Capparis spinosa* pollen was coupled with a viability test at the same RH. The highest viability was found in pollen grains kept at 15% RH. Nevertheless, a low RH is not always the best to maintain viability for long periods. *Chamaerops humilis* pollen grains kept for 144 h at 98% RH have a viability of 58%, whereas those kept for the same period at 15% RH have a viability of only 35% (PACINI & al. 1988).

MATTSSON & al. (1974) have proposed that the stigma pellicle and cuticle regulate water passage towards the grain, at least in the dry stigma type, so that hydration does not occur in very distant species crosses (cf. also KNOX & al. 1976). In the experiment reported in Table 3 we note that even if pollen grains pollinate the stigma of a distant species they hydrate, but their volume may be higher or lower than after a compatible cross. In the intrageneric pollination of *Populus*, most pollen grains from species belonging to sect. *Leuce* fail to adhere to the stigma of species of sect. *Aigeiros*. In the reciprocal crosses pollen tubes are emitted, but they are arrested in the style (GAGET & al. 1984). Furthermore, pollen grains do not hydrate in some cases of heteromorphic selfincompatibility (DULBERGER 1987).

Quite recently it was found that pollen grains exposed to ants lose their viability and do not hydrate because the insect secretes a substance which affects the pollen plasmamembrane (BEATTIE & al. 1984).

The abiotic part of the environment can modify, to a certain extent, the incompatibility reaction. High RHs can overcome the selfincompatibility response (CARTER & MC NEILLY 1976, OCKENDON 1978, KNOX 1984), influencing pollen grains still in the air or on the stigma. In both cases hydration is independent of the stigma or stigmatic inhibitors or recognition substances are diluted (ZUBERI & DICKINSON 1985).

Pollen and spores react with physical recognition to the environment, either during dispersal or on landing. Further physical recognition involving temperature and the quality of the light precedes germination in *Pteridophyta* and chemical recognition starts during the hydration of *Spermatophyta* pollen (KNOX 1984, PETTITT 1985) because the emission of pollen wall protein recognition substances.

A progressive shifting towards chemical recognition occurs with the evolution of land plant reproductive structures. However, if the reproductive structures are passively or actively transported, they can not escape the negative influences of the environment. Morphological and physiological devices must develop to reduce the rate of these environmental stresses.

References

BARNABÁS, B., 1985: Effect of water on germination ability of maize (*Zea mays* L.) pollen. − Ann. Bot. **55**: 201 − 204.
− FRIDVALSKY, L., 1984: Adhesion and germination of differently treated maize pollen grains on the stigma. − Acta Bot. Hungarica **30**: 329 − 332.
BEATTIE, A. J., TURNBULL, C., HOUGH, T., JOBSON, S., KNOX, R. B., 1984: The vulnerability of pollen and fungal spores to ant secretions: evidence and some evolutionary implications. − Amer. J. Bot. **72**: 606 − 614.
BLACKMORE, S., BARNES, S. H., 1986: Harmomegathic mechanisms in pollen grains. − In BLACKMORE, S., FERGUSON, I. K., (Eds.): Pollen and spores: form and function. − Linn. Soc. Symp. Ser. **12**: 137 − 149.
BOWER, F. O., 1935: Primitive land plants. − London: Macmillan.
BUCHEN, B., SIEVERS, A., 1981: Sporogenesis and pollen grain formation. − In KIERMAYER, O., (Ed.): Cytomorphogenesis in plants, pp. 349 − 376. − Wien, New York: Springer [ALFERT, M., & al., (Eds.): Cell biology monographs **8**.]
CARTER, A. L., MCNEILLY, T., 1976: Increased atmospheric humidity post-pollination: a possible aid to the production of inbred line seeds from mature flowers in the Brussels sprout (*Brassica oleracea* var. *gemmifera*). − Euphytica **25**: 532 − 536.
DUCKER, S. C., PETTITT, J. M., KNOX, R. B., 1978: Biology of Australian seagrasses: pollen development and submarine pollination in *Amphibolis antarctica* and *Thalassodendron ciliatum* (*Cymodoceaceae*). − Austral. J. Bot. **26**: 265 − 285.
DUHOUX, E., 1982: Mechanism of exine rupture in hydrated taxoid type of pollen. − Grana **21**: 1 − 7.
DULBERGER, R., 1987: Fine structure and cytochemistry of stigma surface and incompatibility in some distylous *Linum* species. − Ann. Bot. **59**: 203 − 217.
DUMAS, C., GAUDE, C., 1983: Stigma-pollen recognition and pollen hydration. − Phytomorphology **31**: 191 − 201.
− KNOX, R. B., MCCONCHIE, C. A., RUSSEL, S. D., 1984: Emerging physiological concepts in fertilization. − What's New in Plant Physiology **15**: 17 − 20.
FERRARI, T. E., BEST, V., MORE, T. A., COMSTOCK, P., MUHAMMAD, A., WALLACE, D. H., 1985: Intercellular adhesions in the pollen-stigma system: pollen capture, grain binding, and tube attachments. − Amer. J. Bot. **72**: 1466 − 1474.
GAGET, M., SAID, C., DUMAS, C., KNOX, R. B., 1984: Pollen-pistil interactions in interspecific crosses of *Populus* (sections *Aigeiros* and *Leuce*): pollen adhesion, hydration and callose responses. − J. Cell Sci. **72**: 173 − 184.
GILISSEN, L. J. W., 1977: The influence of the relative humidity on the swelling of pollen grains in vitro. − Planta **136**: 299 − 301.
HESLOP-HARRISON, J., 1979 a: Pollen walls as adaptive systems. − Ann. Missouri Bot. Gard. **66**: 813 − 829.
− 1979 b: An interpretation of the hydrodynamics of pollen. − Am. J. Bot. **66**: 737 − 743.
− HESLOP-HARRISON, Y., 1982: The specialized cuticles of the receptive surface of angiosperm stigmas. − In CUTLER, D. F., ALVIN, K. L., PRICE, C. E., (Eds.): The plant cuticle, pp. 99 − 119. − London: Academic Press.

Heslop-Harrison, J., Heslop-Harrison, Y., 1985 a: Germination of stress tolerant *Eucalyptus* pollen. – J. Cell Sci. **73**: 135–157.

– – 1985 b: The secretory system of *Oenothera organensis* Munz.: some development and quantitative studies. – Israel J. Bot. **34**: 187–204.

– – Shivanna, K. R., 1984: The evaluation of pollen quality and further appraisal of the fluorochromatic (FCR) test procedure. – Theor. Appl. Genet. **67**: 367–375.

– Heslop-Harrison, J. S., Heslop-Harrison, Y., 1986: The compartment of the vegetative nucleus and generative cell in the pollen and pollen tubes of *Helleborus foetidus* L. – Ann. Bot. **58**: 1–12.

Heslop-Harrison, J. S., Heslop-Harrison, Y., Reger, B. J., 1987: Anther-filament extension in *Lilium:* potassium ion movement and some anatomical features. – Ann. Bot. **59**: 505–515.

Heslop-Harrison, Y., Shivanna, K. R., 1977: The receptive surface of the angiosperm stigma. – Ann. Bot. **41**: 1233–1258.

– Heslop-Harrison, J. S., Heslop-Harrison, J., 1986: Germination of *Corylus avellana* L. (hazel) pollen: hydration and the function of the oncus. – Acta Bot. Neerl. **35**: 265–284.

Johnson, D. M., 1985: New records for longevity of *Marsilea sporocarps*. – Amer. Fern. J. **75**: 30–31.

Kaul, V., Theunis, C. H., Palser, B. F., Knox, R. B., Williams, E. G., 1987: Association of the generative cell and vegetative nucleus in pollen tubes of *Rhododendron*. – Ann. Bot. **59**: 227–235.

Kerhoas, C., Gay, G., Dumas, C., 1987: A multidisciplinary approach to the study of the plasma membrane of *Zea mays* pollen during controlled dehydration. – Planta **171**: 1–10.

Knox, R. B., 1984: Pollen-pistil interactions. – In Linskens, H. F., Heslop-Harrison, J., (Eds.): Cellular interactions, pp. 508–608. – Berlin, Heidelberg, New York: Springer [Pirson, A., Zimmermann, M. H., (Eds.): Encyclopedia of plant physiology **17**.]

– Clarke, A. E., Harrison, S., Smith, P., Marchalonis, J. J., 1976: Cell recognition in plants. determinants of the stigma surface and their pollen interactions. – Proc. Natl. Acad. Sci. U.S.A. **73**: 2788–2792.

Kress, W. J., 1986: Exineless pollen structure and pollination systems of tropical *Heliconia* (*Heliconiaceae*). – In Blackmore, S., Ferguson, I. K., (Eds.): Pollen and spores: form and function. – Linn. Soc. Symp. Ser. **12**: 329–345.

Luza, J. G., Polito, V. S., 1987: Effects of dessication and controlled rehydration on germination in vitro of pollen of walnut (*Juglans* spp.). – Plant Cell Environ. **10**: 487–492.

Machlis, L., Rawitscher-Kunkel, E., 1967: The hydrated megaspore of *Marsilea vestita*. – Amer. J. Bot. **54**: 689–694.

Marginson, R., Sedgley, M., Knox, R. B., 1985: Physiology of post-pollination exudate production in *Acacia*. – J. Exper. Bot. **36**: 1660–1668.

Mattsson, O., Knox, R. B., Heslop-Harrison, J., Heslop-Harrison, Y., 1974: Protein pellicle of stigmatic papillae as a probable recognition site in incompatibility reactions. – Nature **247**: 298–300.

Miller, J. H., 1980: Differences in the apparent permeability of spore walls and protallial cell walls in *Onoclea sensibilis*. – Amer. Fern. J. **4**: 119–123.

Niklas, K. J., 1985: The aerodynamics of wind pollination. – Bot. Rev. **51**: 328–386.

Ockendon, D. J., 1978: Effects of exine and humidity on self-incompatibility in *Brassica oleraceae*. – Theor. Appl. Genet. **52**: 113–117.

Pacini, E., 1986: An approach to harmomegathy. – In Cresti, M., Dallai, R., (Eds.): Biology of reproduction and cell motility in plants and animals. – Siena: Siena University.

- BASSANI, M., FRANCHI, G. G., 1988: Angiosperm pollen viability and volume after exposure to different relative humidities. − In KNOX, R. B., SINGH, M. B., TROIANI, L. F., (Eds.): Pollination 1988, pp. 160−164. − Melbourne: University of Melbourne.
- FRANCHI, G. G., HESSE, M., 1985: The tapetum: its form, function, and possible phylogeny in *Embryophyta*. − Pl. Syst. Evol. **149**: 155−185.
- PETTITT, J. M., 1979: Ultrastructure and cytochemistry of spore wall morphogenesis. − In DYERS, A. F., (Ed.) The experimental biology of ferns, pp. 213−252. − London: Academic Press.
- 1985: Pollen tube development and characteristics of the protein emission in conifers. − Ann. Bot. **5**: 379−397.
- PLATT-ALOIA, K. A., LORD, E. M., DE MASON, D. A., THOMSON, W. W., 1986: Freeze-fracture observations on membranes of dry and hydrates pollen from *Collomia, Phoenix* and *Zea*. − Planta **168**: 291−298.
- SHIVANNA, K. R., HESLOP-HARRISON, J., HESLOP-HARRISON, Y., 1983: Heterostyly in *Primula*. 3. Pollen water economy: a factor in the intramorph-incompatibility response. − Protoplasma **117**: 175−184.
- JOHRI, B. M., 1985: The angiosperm pollen: structure and function. − New Delhi: Wiley Eastern.
- SINGH, H., 1978: Embryology of gymnosperms. − In ZIMMERMAN, W., & al., (Eds.): Encyclopedia of plant anatomy 10/2. − Berlin: Gebrüder Borntraeger.
- STEAD, A. D., ROBERTS, I. N., DICKINSON, H. G., 1979: Pollen-pistil interactions in *Brassica oleracea:* events prior to pollen germination. − Planta **146**: 211−216.
- − − 1980: Pollen-stigma interaction in *Brassica oleracea:* the role of stigmatic proteins in pollen grain adhesion. − J. Cell Sci. **42**: 417−423.
- TAYLOR, W. C., LUEBKE, N. T., 1986: Germinating spores and growing sporelings of acquatic *Isoetes*. − Amer. Fern J. **76**: 21−24.
- THANIKAIMONI, G., 1978: Pollen morphological terms: proposed definitions − 1. − 4th Int. Palynol. Conf., Lucknow (1976−77) **1**: 228−239.
- 1986: Pollen apertures: form and function. − In BLACKMORE, S., FERGUSON, I. K., (Eds.): Pollen and spores: form and function. − Linn. Soc. Symp. Ser. **12**: 119−136.
- THEUNIS, C. H., MCCONCHIE, C. A., KNOX, R. B., 1985: Three-dimensional reconstruction of the generative cell and its wall connection in mature bicellular pollen of *Rhododendron*. − Micron Microscopica Acta (HESLOP-HARRISON Festschrift) **16**: 225−231.
- WESTGATE, M. E., BOYER, J. S., 1986: Silk and pollen water potentials in maize. − Crop Sci. **26**: 947−956.
- WILLEMSE, M. T. M., REZNICKOVA, S. A., 1980: Formation of pollen in the anther of *Lilium*. 1. Development of the pollen wall. − Aca Bot. Neerl. **29**: 127−140.
- WODEHOUSE, R. P., 1935: Pollen grains. − New York: Mc Graw-Hill.
- ZIEGLER, H., 1959: Über die Zusammensetzung des „Bestäubungstropfens" und den Mechanismus seiner Sekretion. − Planta **52**: 587−599.
- ZUBERI, M. I., DICKINSON, H. G., 1985: Pollen-stigma interaction in *Brassica*. 3. Hydration of pollen grains. − J. Cell Sci. **26**: 321−336.

Address of the author: Prof. Dr E. PACINI, Dept. of Environmental Biology, Botanical Section, University of Siena, I-53100 Siena, Italy.

Pl. Syst. Evol. [Suppl. 5], 71–79 (1990)

Fern spores:
evolutionary levels and ecological differentiation

ALICE F. TRYON

Received October 2, 1987; in revised form April 15, 1989

Key words: Pteridophytes, ferns. – Spores, ecology, evolution.

Abstract: The development of exospore relative to the perispore and type of aperture are used to assess evolutionary levels of spores in the major families of ferns. Spores of primitive families usually have trilete aperture and walls of thick exospore overlaid by a thin, conforming perispore; spores of derived families are largely monolete with thin exospore below a complex perispore. Spores at a specialized level have elaborate strata of either primitive or derived type.

Correlation between surface structure of spores and ecology of *Asplenium* and of *Pyrrosia* show epiphytic species have spores with more elaborate surface contours than spores of epilithic or terrestrial. The data suggests that surface contours may reflect the ecological differences to which species are adapted.

The diverse surface and wall structure of fern spores is considered in evolutionary and in ecological perspective, based on an electron microscope review of all genera. The general evolutionary relations of spores proposed here, at the level of family, provide a limited review of the data but supply a framework for more detailed assessment of pteridophyte diversity. The main components and ontogeny of the sporoderm in fern spores are known largely from studies of sporogenesis in *Ophioglossum, Osmunda* and *Blechnum* (LUGARDON 1971). The initial and main part of the wall, consisting of sheaths or feuillets that are enveloped by amorphous sporopollenin, forms the exospore. This is overlaid by perispore, the outer part of the wall, derived from tapetal material. A cellulosic endospore, adjacent to the protoplast, is evident in fern spores after germination.

Evolutionary levels

Three general evolutionary levels are based on the relative development of the exospore and perispore, and to some extent on the trilete or monolete type of aperture, in selected families as diagrammed in Fig. 1. The position of the families is based on correlation of the spores with other characteristics by which they are usually regarded as primitive or derived. Spores of *Hymenophyllaceae, Grammitidaceae,* and several small families that are essentially similar to included ones are omitted from the chart.

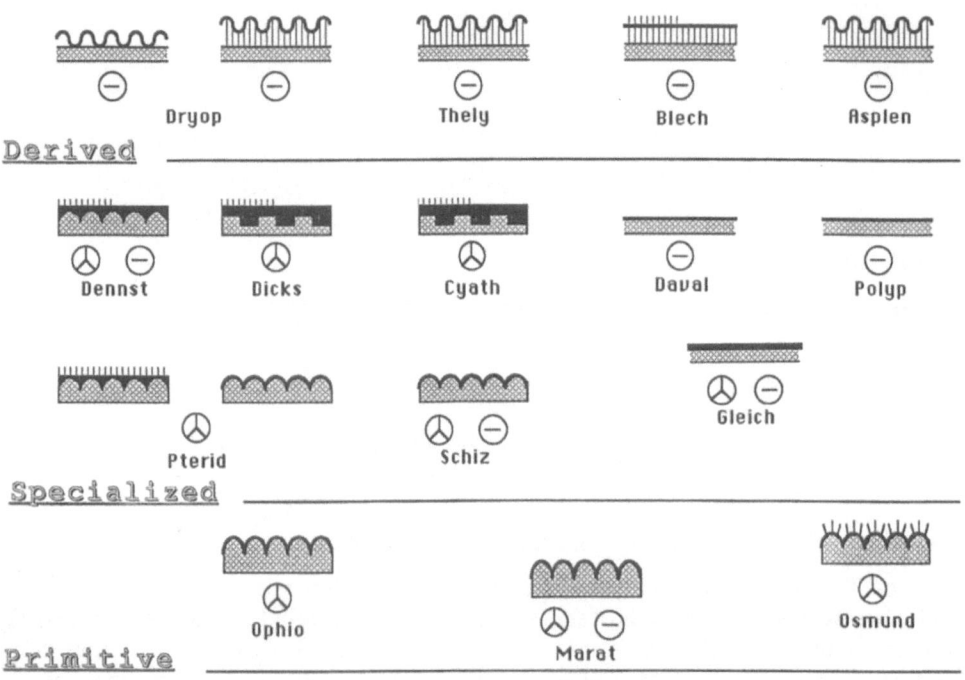

Fig. 1. Evolutionary levels of wall structure in fern spores. Aperture: trilete, three radii in circle, monolete, one bar within circle; Exospore: hatched lower layer; Perispore: black upper layer, undulate folds, or radiant lines on outer surface, rodlets: vertical lines on outer surface, Inner perispore: pillared lines below surface. Family abbreviations: ASPLENiaceae, BLECHnaceae, CYATHeaceae, DAVALleaceae, DENNSTaediaceae, DICKSoniaceae, DRYOPteridaceae, GLEICHeniaceae, MARATtiaceae, OPHIOglossaceae, OSMUNDaceae, POLYPodiaceae, PTERIDaceae, SCHIZaeaceae, THELYpteridaceae

Figs. 2 – 7 and 9 – 11. Surface and wall sections of fern spores; Fig. 8 pollen wall. Collections cited are in the Harvard University Herbaria, except as noted. – Fig. 2. *Ophioglossum engelmannii* PR., trilete, Mexico, REEDER & al. 3401, × 1 000. – Fig. 3. *O. lusitanicum* L., complex exospore strata below thin perispore (arrow), France, LUGARDON 639 (TL), × 10 000. – Fig. 4. *Macroglossum alidae* COPEL., exospore tubercles below abraded perispore, Sarawak, MOLESWORTH-ALLEN 3197, × 10 000. – Fig. 5. *Osmunda cinnamomea* L., surface detail, echinate perispore on coarse exospore tubercles, Massachusetts, DEAN in 1886, × 5 000. – Fig. 6. *Dryopteris filix-mas* L., wall section, inflated perispore fold, Germany, REICHSTEIN 3164, × 5 000. – Fig. 7. *Thelypteris concinna* (WILLD.) CHING, exospore (below arrow), pillared perispore above, Costa Rica, SKUTCH 2597, × 10 000. – Fig. 8. *Statice sinuata* L. [*Limonium sinuatum* (L.) MILL.], columellate exine of angiosperm pollen, Morocco, MAIRE & WILCZEK 1048, (NOWICKE & SKVARLA 1979). – Fig. 9. *Gleichenia dicarpa* R. BR., proximal face (left) with special flange; distal face (right), flange at arrow, New South Wales, Australia, TRYON & TRYON 7374, × 1 000. – Fig. 10. *Anemia underwoodii* MAXON, trilete, echinate, Cuba, MORTON 3397, × 500. – Fig. 11. *Cheilanthes guanchica* BOLLE, perispore with more or less coalescent rodlets below outer perispore, France, REICHSTEIN 4122, × 10 000. – Fig. 12. *Sphaeropteris elongata* (HOOK.) TRYON, wall section, granulate perispore above pitted exospore, Columbia, PENNELL & KILLIP 6304, × 5 000

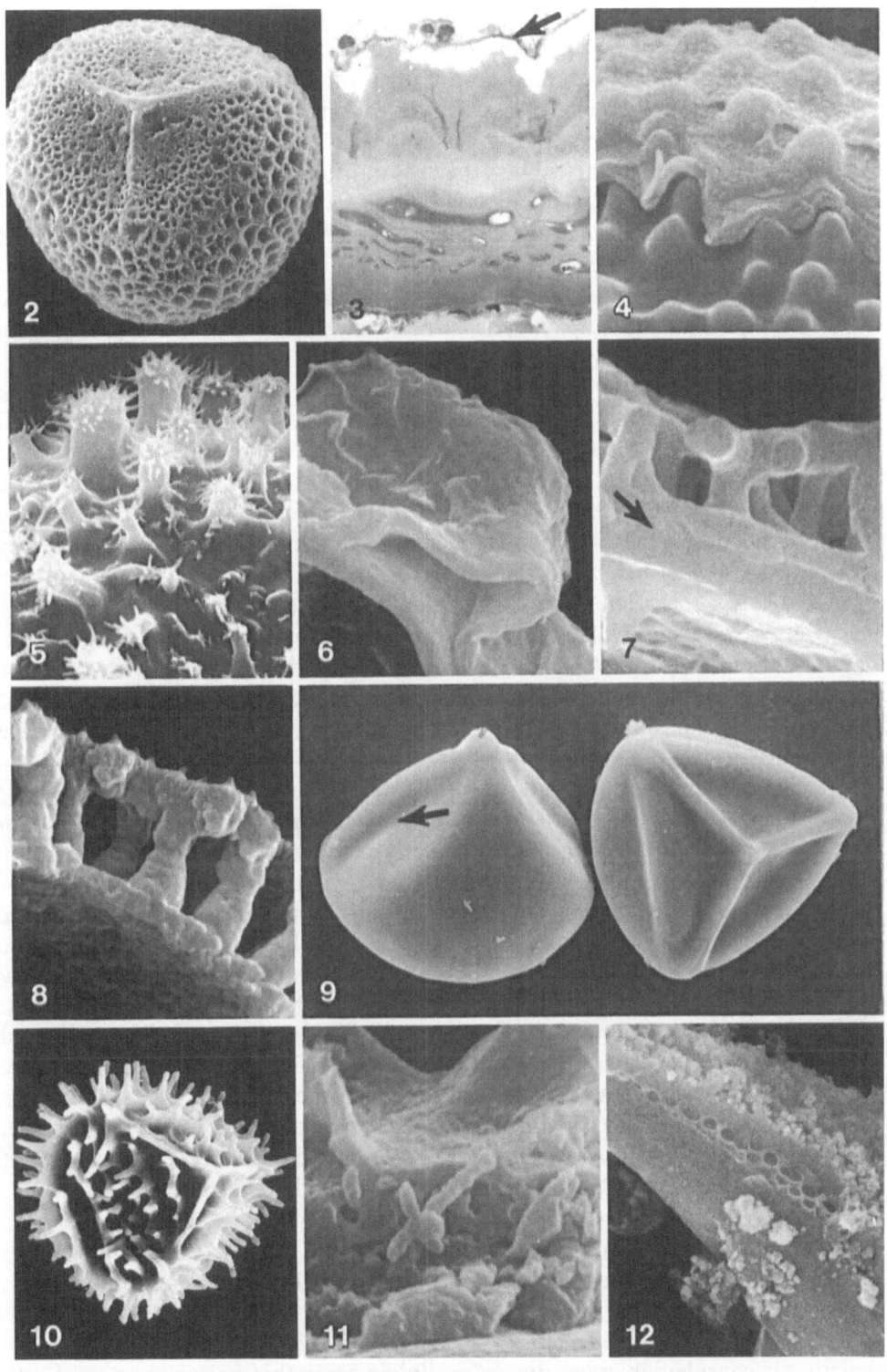

The spore shape and type of aperture reflects the alignment of spores in the tetrad and the position of the tetraspores during sporogenesis. Trilete spores, tetrahedrally arranged, are regarded as primitive for they occur early in the fossil record, and predominate in the less derived families. Monolete spores are most common in the derived ferns and the tetraspores are decussately arranged in these with one pair perpendicular to the other. Both trilete and monolete spores occur in eight of the 25 filicalean families recognized in TRYON & TRYON (1982) including the *Marattiaceae, Gleicheniaceae, Schizaeaceae, Vittariaceae, Dennstaedtiaceae*. They are rarely monolete in the *Pteridaceae*, and rarely trilete in the *Polypodiaceae* and *Thelypteridaceae*. In spite of these variants the type of aperture, considered along with the wall structure, supplies useful evidence relative to the primitive or derived state of spores.

Primitive wall structure (Figs. 2 – 5). The main contours of spores in the most primitive families, *Marattiaceae, Ophioglossaceae*, and *Osmundaceae*, are formed by thick exospore overlaid by a thin, conforming perispore. The surface may be coarsely rugate, reticulate, echinate, or tuberculate (Figs. 2 – 4). Walls of this type in spores of the Carboniferous genus *Scolecopteris*, of the *Marattiaceae* (MILLAY 1979), also indicate the primitive state of this structure. The spores are usually trilete, as in *Ophioglossum engelmannii* (Fig. 2), except in a few genera of the *Marattiaceae*. Tuberculate spores with echinulate perispore in *Osmunda cinnamomea* (Fig. 5) characterize all genera of the *Osmundaceae*, and are unique in the pteridophytes. Spores of the *Hymenophyllaceae* are consistently trilete, with exospore elaborated in three strata, below a thin perispore, indicating a primitive type of structure. The endospore may be well developed as the spores often germinate precociously within the spore wall.

Derived wall structure (Figs. 6, 7, and 13 – 18). The main contours and largest part of the wall is formed by an elaborate perispore that overlays a relatively thin exospore. These contrast with the primitive type in which the exospore forms the most prominent part of the wall. The *Dryopteridaceae, Thelypteridaceae, Blechnaceae*, and *Aspleniaceae* have spores of this structure. The extent of perispore development varies in these families, but usually involves differences in thickness and number of layers. Folds, formed by the outer layer of perispore, are variously inflated (Figs. 6 and 13), compressed (Fig. 14), extended in echinae (Figs. 16 and 17), fenestrate (Fig. 15), or reticulate (Fig. 18). The inner perispore is often pillared as in *Thelypteris concinna* (Fig. 7) and *Asplenium serra* (Fig. 17). This structure resembles the columellate wall characteristic of angiosperm pollen (Fig. 8), NOWICKE & SKVARLA (1979). However, in pollen the columellae are formed in the exine, while the pillared structure of fern spores is within the perispore. Comparisons of exine in angiosperm and gymnosperm pollen with the exospore in fern spores (LUGARDON 1978) show basic structural and ontogenetic differences in the walls of pollen and spores. Although columellate pollen and pillared walls of fern spores clearly differ in development and structure they probably have common functions of mechanical support or storage.

Specialized wall structure (Figs. 9 – 12 and 19 – 24). In these families spore structure is considerably more diverse including some that have spores similar to those at the primitive or derived level, but either the exospore or perispore is more specialized. The major contours may be formed by exospore with thin, conforming perispore, as in primitive spores, or the perispore may be especially elaborated.

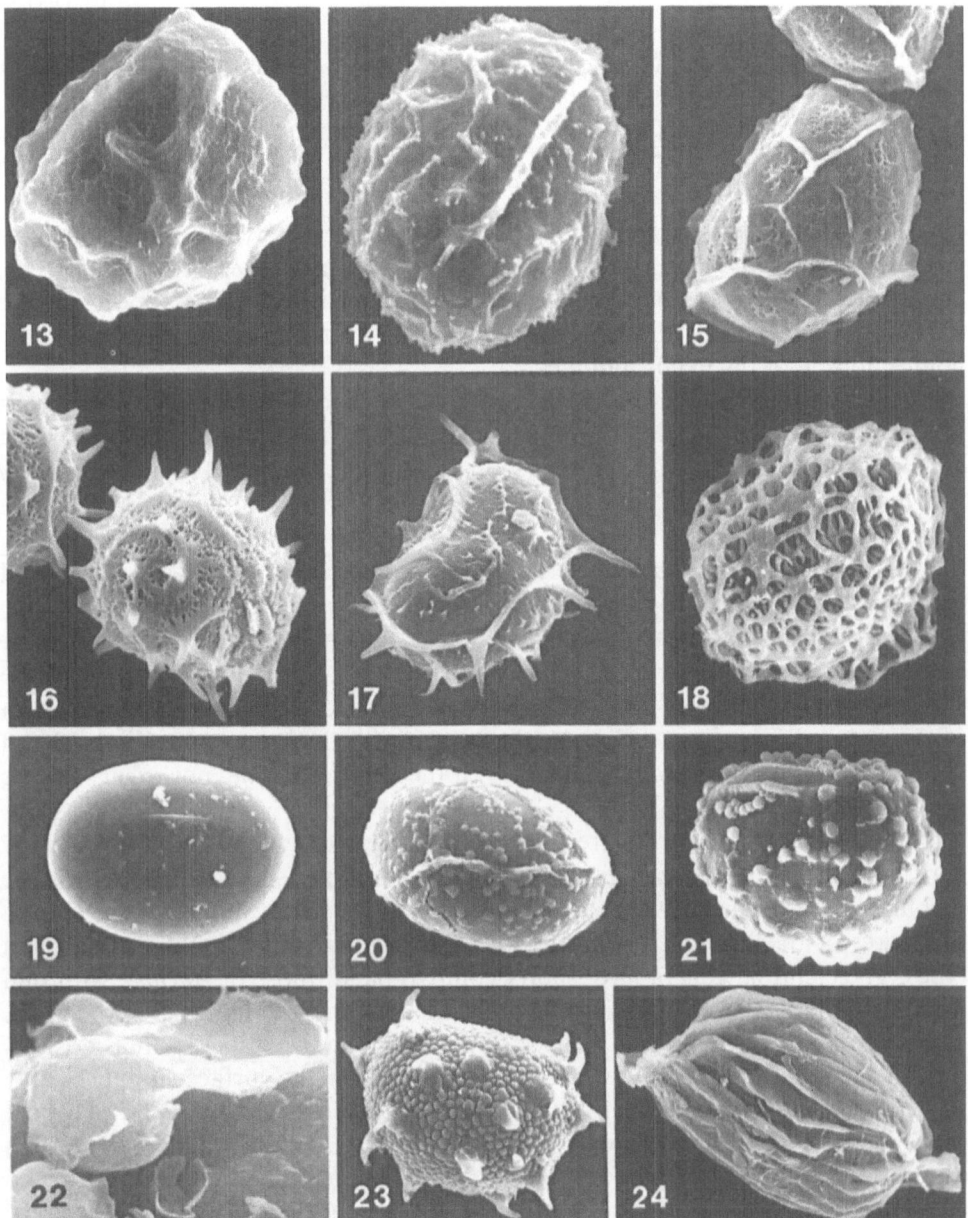

Figs. 13 – 24. Fern spore diversity correlated with ecology. – Figs. 13 – 18. *Asplenium* spores,
× 1 000. – Terrestrial: 13. *A. castaneum* SCHLECHT. & CHAM., coarse, low folds, Mexico,
PRINGLE 6150. – Lithophytic: 14. *A. ruta-muraria* L., folds more or less compressed, England, A. TRYON, in 1977. – Terrestrial or epiphytic: 15. *A. auritum* SW., fenestrate, low
wings, St. Kitts, PROCTOR 19537. – Epiphytic: 16. *A. falcinellum* MAXON, fenestrate, echinate, Costa Rica, MADISON 733. – 17. *A. reptans* HOOK., echinate, Ecuador, SPRUCE
5336. – 18. *A. serra* LANGSD. & FISCH., reticulate with pillared perispore below, Dominican
Republic, GASTONY & al. 638. – Figs. 19 – 24. *Pyrrosia* spores, × 500. – Epiphytic or epilithic: 19. *P. costata* (PR.) TAGAWA & IWATSUKI, smooth proximal face, short aperture
above center, Burma, KINGDON-WARD 20455. – 20. *P. manii* (GIESENH.) CHING, colliculate,
Nepal, HARA & al. 630493. – Epilithic: 21 and 22. *P. petiolosa* (CHRIST) CHING, Anhwei,
China, FAN & LI. 1 – 21. Globulose surface. – 22. Globules below thin perispore,
× 5 000. – Usually epiphytic: 23. *P. piloselloides* (L.) PRICE, verrucate, echinate, Hainan,
China, LIANG 64275. – 24. *P. angustata* (SW.) CHING, wrinkled, Malacca, CUMING 372

The type of aperture is not consistent, as both trilete and monolete spores occur in several of these families (Fig. 1).

Elaboration of the surface or wall structure common to some families seems to reflect alliances. Spores of the *Gleicheniaceae* have an exospore flange, within the proximal portion, that occurs in both trilete and monolete spores. This is characteristic of both *Dicranopteris* and *Gleichenia* (Fig. 9), and also occurs in spores of the *Stromatopteridaceae, Dipteridaceae, Cheiropleuriaceae,* and *Matoniaceae,* suggesting relationships of these families with the *Gleicheniaceae.* Spores of the Triassic *Plebopteris smithii* (Ash & al. 1982) have a similar flange that may imply connections with the extant families.

Genera of the *Schizaeaceae* have exceptionally diverse spores. The exospore is usually less developed than in primitive types, and the aperture is trilete, except for monolete *Schizaea* spores. The perispore is elaborate in some species of *Lygodium* in which a grid-like structure incorporating silica, forms a large part of the wall. In *Anemia* coarse, parallel ridges or spines (Fig. 10) are overlaid by a thin perispore. Spores comparable to *Anemia* are known in allied genera from Cretaceous and Early Tertiary deposits in many parts of the world.

Polypodiaceae spores are usually monolete, have a relatively thick exospore, and thin perispore (Figs. 19–22). The wall structure appears closer to that of the *Gleicheniaceae* than other families with specialized spores. However, some genera of the *Polypodiaceae* have more complex spores with thick perispore (Figs. 22 and 23) and these do not support alliances with the *Gleicheniaceae,* nor other families at this level.

Spores of the *Davalliaceae* are consistently monolete with low contours usually derived from the exospore. This is overlaid by a thin perispore with some surface globules, and superficially appears similar to *Polypodiaceae* spores.

Spores of the *Pteridaceae*, with rare exception, are trilete, and of two main types. The taenitoid genera have spores comparable to those of *Anemia*, with thin perispore conforming to a thick exospore, and an equatorial flange (Tryon 1985). Those of the cheilanthoid genera have thin exospore and the perispore is cristate, reticulate or echinate usually with rodlets on or below the surface (Fig. 11).

Spores of the *Dicksoniaceae* and *Cyatheaceae* are consistently trilete. Exospore pits (Fig. 12), common to several but not all genera in both families, are often obscured by granulate perispore. These pits are especially well developed and characterize *Cnemidaria* of the *Cyatheaceae.* Rodlets, usually short and cylindrical with a clear core, may fuse in longer strands or laminate forms. These are common to several genera of the *Cyatheaceae* and *Dicksoniaceae* (Gastony 1979, Tryon & Tryon 1982), and some genera of the *Dennstaedtiaceae, Pteridaceae,* and *Blechnaceae,* as well as a few of the *Schizaeaceae* and *Gleicheniaceae.* They are not developed in spores of the primitive families, nor in the *Polypodiaceae, Thelypteridaceae,* or *Dryopteridaceae.*

Spores of the *Dennstaedtiaceae* often have the perispore elaborated into several strata and thicker than the exospore. The perispore varies from a simple granulate surface in *Pteridium*, or rodlets in *Microlepia*, to more elaborate echinate or reticulate forms in *Hypolepis* and *Blotiella*. The aperture also varies within genera. Both types of aperture, as well as intermediate forms occur in the same collection of *Lindsaea pallida* (Tryon 1985).

Spores of the *Grammitidaceae* are usually trilete, with thin exospore and perispore. The wall structure is complex due to precocious germination within intact spores, and the early development of endospore and a special pseudo-endospore formation, adjacent to the spore protoplasm.

Ecological change

Surface contours of spores in several genera correlate to some extent with the ecology of the species. This is especially clear in *Asplenium* in which species often are ecologically recognized as terrestrial, epilithic or epiphytic. Four main spore types – folded, fenestrate, echinate, and reticulate – have been distinguished in a review of 111 species including a diverse systematic and geographic sample of species (Fig. 25). The echinate and reticulate spores (Figs. 16 – 18) are regarded as more derived than folded or fenestrate ones (Figs. 13 – 15). Epiphytic species have either echinate or reticulate spores in 46% of the species. The epilithic aspleniums have folded or fenestrate spores in 95% of the species, and the terrestrial species have folded or fenestrate spores in 85% of the species while only 15% have spores of the more derived form (Fig. 25).

Correlation of surface contours and ecology of the species of *Pyrrosia* (*Polypodiaceae*) is based on the monograph of 44 species of the genus (HOVENKAMP 1986). The species are predominantly epiphytic but some are epilithic, or ecologically diverse. Five spore types recognized in *Pyrrosia* by VAN UFFELEN & HENNIPMAN (1985) are analysed here. They review variation within the five types, each named for a representative species in their study. Descriptive names for their types are applied here, based on the predominant form of surface, include: smooth (Fig. 19), colliculose (Fig. 20), globulose (Figs. 21 and 22), wrinkled (Fig. 24), and verrucose (Fig. 23). Spores with wrinkled or verrucose surfaces are considered to represent more elaborate, derived forms (Fig. 26). Among the exclusively epilithic species,

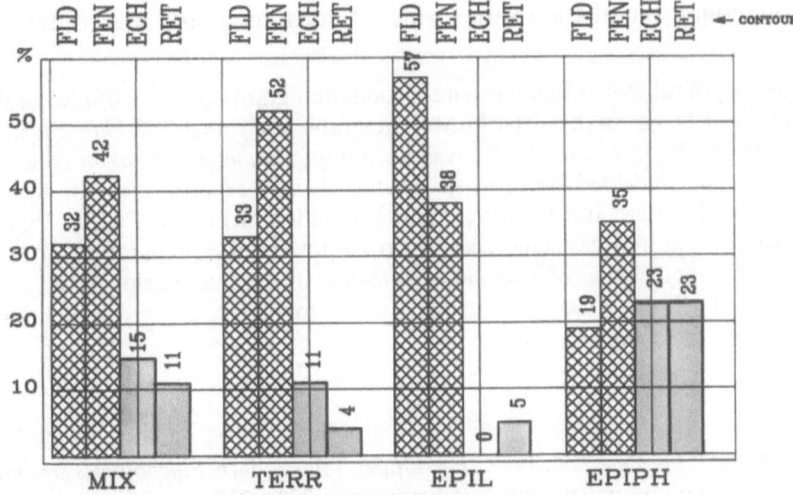

Fig. 25. Correlation of ecology of 111 species of *Asplenium* with surface contours of the spores; surfaces FLD folded, FENestrate, ECHinate, RETiculate. Percentage of different spore types according to the ecology of the species – MIXed ecology, TERRestrial, EPI-Lithic, EPIPHytic

78   A. F. Tryon:

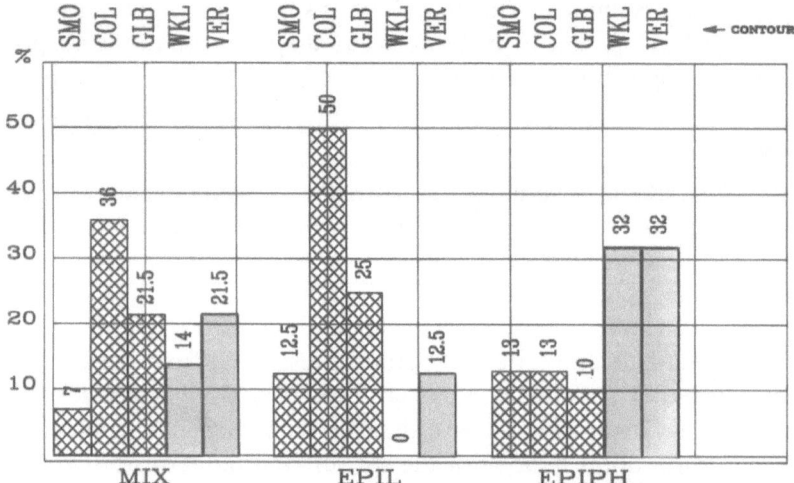

Fig. 26. Correlation of ecology of 44 species of *Pyrrosia* with surface contours of the spores; surfaces SMOoth, COLliculose, GLB globulose, WKL wrinkled, or VERrucose. Percentage of different spore types according to ecology of the species – MIXed ecology, EPILithic, EPIPHytic

87.5% have spores that are either smooth, colliculose or globulose and only 12.5% have spores with a specialized verrucate surface. Spores of 64% of the epiphytic species have more specialized wrinkled or verrucose spores, and 36% have simpler types of surface. Spores of 64.5% of the species with both epilithic and epiphytic ecology have smooth, colliculose or globulose surfaces, while 35.5% have more specialized wrinkled or verrucose spores. Thus, in *Pyrrosia* as well as in *Asplenium*, correlation between the obligate epiphytic habit and elaboration of the spores suggest that surface complexity may relate to ecological specialization of the species. However the functional role of the spore surface is not known and may differ in these genera.

This work was supported by National Science Foundation grants DEB 81-05726, BSR 84-07046, to Rolla and Alice Tryon. The TEM photograph of *Ophioglossum lusitanicum* (Fig. 3), was supplied by B. Lugardon. The wall section of *Limonium sinuatum* (Fig. 8) by J. W. Nowicke. I am indebted to Edward Seling for the excellence of the SEM micrographs, to Lloyd Schoenbach for the computer rendition of Figs. 1, 25, and 26, to Phyleen Stewart for editorial assistance, and Lydia Vickers for preparation of the plates. I am especially appreciative of comments in review of the manuscript by B. Lugardon, A. H. Knoll, K. U. Kramer, and particularly to Rolla Tryon for helpful suggestions throughout the spore studies.

References

Ash, S., Letwin, R. A., Traverse, A., 1982: The Upper Triassic fern *Phlebopteris smithii* (Daugherty) Arnold and its spores. – Palynology **6**: 203–219.
Gastony, G. J., 1979: Spore morphology in the *Cyatheaceae* 3. The genus *Trichipteris*. – Am. J. Bot. **66**: 1238–1260.
Hovenkamp, P., 1986: A monograph of the fern genus *Pyrrosia*. – Leiden Bot. Ser. **9**: 1–280.

LUGARDON, B., 1971: Contribution à la connaissance de la morphogenése et de la structure des parois sporales chez les filicinées isosporées. − Thèse 458. L'Univ. Paul Sabatier de Toulouse.

− 1978: Comparison between pollen and pteridophyte spore walls. − 4th Int. Palynol. Confer. Lucknow (1976−1977) **1**: 199−206.

MILLAY, M. A., 1979: Studies of Paleozoic marattialeans: a monograph of *Scolecopteris*. − Palaeontographica **169**: 1−169.

NOWICKE, J. W., SKVARLA, J. J., 1979: Pollen morphology: the potential influence in higher order systematics. − Ann. Missouri Bot. Gard. **66**: 633−700.

TRYON, A. F., 1985: Stasis, diversity and function in spores based on an electron microscope survey of the *Pteridophyta*. − In BLACKMORE, S., FERGUSON, I. K., (Eds.): Pollen and spores: form and function. − Linn. Soc. Symp. Ser. **12**: 233−249.

TRYON, R. M., TRYON, A. F, 1982: Ferns and allied plants with special reference to tropical America. − New York, Berlin, Heidelberg: Springer.

VAN UFFELEN, G., HENNIPMAN, E., 1985: The spores of *Pyrrosia* MIRABEL (*Polypodiaceae*), a SEM study. − Pollen & Spores **27**: 155−198.

Authors' address: ALICE F. TRYON, Harvard University Herbarium, Cambridge, Massachusetts, U.S.A. − New permanent address: Department of Biology, University of South Florida, Tampa, FL 33620-5150, U.S.A.

Pl. Syst. Evol. [Suppl. 5], 81–90 (1990)

The genus *Acacia* (*Leguminosae, Mimosoideae*): its affinities as borne out by its pollen characters

P. GUINET

Received September 30, 1987

Key words: Angiosperms, *Leguminosae, Mimosoideae, Acacia.* – Pollen grain morphology, generic relationships.

Abstract: Possible affinities of the genus *Acacia* s.l. are discussed. Notwithstanding the well-known common characters of the genus with the tribe *Ingeae*, the pollen characters rather suggest its derivation from the *Mimoseae*. The two main pollen groups recognized in *Acacia* do not appear directly related. The results support the hypothesis of pleiophily put forward by PEDLEY in 1986, restricted however to two groups; the first limited to the species included in the genus *Acacia* sensu PEDLEY (*Acacia* subgen. *Acacia*) and the second including the two other subgg. *Aculeiferum* and *Phyllodineae*.

The genus *Acacia*, as most frequently understood, alone constitutes the tribe *Acacieae* of subfam. *Mimosoideae*. BENTHAM (1875) has established the basic outlines of the system for this subfamily, and since then, the tribe has been most generally taken as such, except for the later recognition of the African *Acacia albida* as distinct genus (*Faidherbia*). Although some taxonomists have contended that the genus is far too diverse and difficult to handle, and have described a number of segregate genera (BRITTON & ROSE 1928), it is still now recognized as a natural unit and BENTHAM's subdivisions have remained the base for all authors who attempted to arrange the species in a single sequence. As a consequence, the relationships between the subdivisions of the genus have been much discussed, but affinities between *Acacia* and the other genera of the *Mimosoideae* were rarely considered. Recently, however, PEDLEY (1986) has recognized three segregate genera instead of the single genus *Acacia* and has discussed their possible derivation.

This paper is concerned with the relationships between *Acacia* s.l. and other mimosoid genera, as they appear from their pollen morphological traits. For that purpose, I have used indifferently either the three subgenera of *Acacia* formally named by VASSAL (1972) which correspond to the three pollen groups I had previously established (GUINET 1964) or the three genera recognized by PEDLEY (1986). The change of taxonomic rank does not change the content of each group. Furthermore, the recently improved generic classification of the *Mimosoideae* (references given below) allow more accurate comparisons than in the past.

Pollen characters in *Acacia* sensu lato

Main characters. No new data necessitate a new discussion of *A. albida* DEL., segregated alone in the genus *Faidherbia* or maintained in *Acacia* as a proper subgenus [subgen. *Faidherbia* (A. CHEV.); MAIRE (1987)]. One should recall that this species has a very unusual combination of characters: stipular spines as in *Acacia* subgen. *Acacia*, flowers with a disc and a gynophore and no floral involucel (but flowers in spikes) as in *Acacia* subgen. *Aculeiferum*, stamens very shortly but regularly united at their bases in a tube (as in most *Ingeae*), phytochemistry (BELL & EVANS 1978; EVANS, pers. comm.) intermediate between *Acacia* subgen. *Aculeiferum* and tribus *Ingeae*, porate polyads with a granular exine structure and very faint areoles, as in many species of *Albizia* (*Ingeae*). When more emphasis is given to some characters, *Faidherbia* could be as well included in the *Ingeae* (GUINET 1969, GUINET & LUGARDON 1976).

Used alone, the pollen characters in *Acacia* (*Faidherbia* excluded) allow its subdivision into the following main groups:

1 Exine with a columellar infratectal structure; polyads distally colporate; pseudocolpi absent; tectum flat, smooth, crossed by broad, generally numerous, perforations

1′ Exine with a granular infratectal structure; polyads distally porate
  2 Tectum smooth, flat or undulated, crossed by narrow channels or imperforate; pseudocolpi absent
  2′ Tectum reticulate (supratectal reticulum), flat, most frequently imperforate; pseudocolpi present or absent

Congruence with the present classification (according to characters listed by PEDLEY 1986) is as follows:

1 Stipular spines present; leaves bipinnate, flowers in spikes or heads, the latter always with an involucel on the peduncle: *Acacia* subgen. *Acacia* (*A.* sect. *Gummiferae* BENTH.), genus *Acacia* PEDLEY (1986)

1′ Stipular spines absent or if present then leaves modified to phyllodes; leaves bipinnate or phyllodes; flowers in spikes or heads, the latter never with an involucel on the peduncle.
  2 Flowers with a disc and ovary on a gynophore; leaves bipinnate; plants with prickles or if not then extrafloral nectaries absent: *Acacia* subgen. *Aculeiferum* VAS. (1972) (*A.* sectt. *Vulgares* and *Filicinae* both of BENTH.), genus *Senegalia* PEDLEY (1986)
  2′ Flowers without a disc and ovary not on a gynophore; leaves modified to phyllodes or if bipinnate then with extrafloral nectaries; plants without prickles: *Acacia* subgen. *Phyllodineae* SER. (combined sectt. *Phyllodineae*, *Botrycephalae* and *Pulchellae*, all of BENTH.), genus *Racosperma* (DC.) MART. (PEDLEY 1986)

The strong correspondence between these independent sets of characters allows the distinction of two groups:

(1) The first, represented by the comparatively small genus *Acacia* sensu PEDLEY with ±270 spp., appears the most clearly defined. With the relocation of *A. willardiana* ROSE (PEDLEY 1975) and *A. coulteri* GRAY ex BENTH. in subgen. *Aculeiferum*, pollen exine structure and apertural type appear the most stable characters in the genus (with no exception).

(2) The second with ±950 spp. includes the subgg. *Aculeiferum* (*Senegalia*) and *Phyllodineae* (*Racosperma*). Their palynological distinction rests on two characters: (a) in subgen. *Aculeiferum* the exine is not suprareticulate, and well differentiated pseudocolpi are absent. (b) In subgen. *Phyllodineae* the exine is most frequently suprareticulate (if not, pseudocolpi are present), and most species have pseudocolpi. The ±100 species that lack pseudocolpi all have a suprareticulate tectum and phyllodes (GUINET 1986).

Other characters. A p e r t u r e s are more than three and non-equatorial in subgen. *Acacia* (*Acacia* sensu PEDLEY), i.e., each grain of the polyad is pluriaperturate. Apertures are pores on the proximal and lateral walls of the grains, but colpori on the distal sides. The colpi never have free ends and are most frequently three in number. Their distribution does not correspond to the symmetry of the grains (four distal sides, in 16 grain polyads, but only three colpori).

In a few species there are four distal colpi but only three with a pore. They belong to the *A. farnesiana* alliance [with *A. caven* MOL., *A. farnesiana* (L.) WILLD., *A. pennatula* (SCHL. & CHAM.) BENTH., *A. pinetorum* HERM., *A. shaffneri* (WATS) HERM., *A. smallii* ISELY, *A. tortuosa* (L.) WILLD.] and also include the two African species *A. erioloba* E. MEY and *A. haematoxylon* WILLD. Most of these may have stamen filaments irregularly connate at the base (no stamen tube), a floral involucel and ± indehiscent pods.

P s e u d o c o l p i (s u b s i d i a r y c o l p i) occur in most species of the Australian subg. *Phyllodineae*, whether they have bipinnate leaves or phyllodes, suggesting that the subgenus has to a large degree evolved independently. In Eastern Australia their differentiation follows a continuous geographic gradient (GUINET 1979, 1986) from the northern tropical phyllodinous species to the species with bipinnate mature leaves. The latter appear to have been directly derived from the Australian species with phyllodes (GUINET 1986, PEDLEY 1978, 1986), and are thus not related to the (bipinnate) species of the other subgenera. However, pseudocolpi have been described in *A. ferruginea* DC., an extra-Australian species of the subgen. *Aculeiferum* (GUINET 1969). In this species they are poorly differentiated and occur only sporadically, and the tectum is smooth.

T e c t a l o r n a m e n t a t i o n a n d c h a r a c t e r i s t i c s. In *Acacia* s.l. the tectum generally forms an even, flat, smooth or reticulate layer. Exceptions are: (a) Very few species of the subgen. *Phyllodineae* have a tectum covered by ± rounded small verrucae, as occur nowhere else in *Acacia* s.l. These species are *A. laccata* PEDL., *A. oraria* PEDL., and *A. melleodora* PEDL., all with pluriverved phyllodes, occur in tropical Northern Queensland and are very isolated (PEDLEY 1978). (b) In subgen. *Aculeiferum* the tectum is not always flat but frequently distinctly undulated, in such a manner that it superficially looks like a "reticulum".

Pollen characters in the tribe *Ingeae*

The relationship between *Acacieae* and *Ingeae* has always been considered as very close, the main distinction being the stamens which are nearly always basally fused into a tube in the *Ingeae*. Generic delimitation in the *Ingeae* follows NIELSEN (1981).

Main characters. Using for this tribe the same pollen characters ordered in the same sequence as in the *Acacieae*, one may recognize the following pollen groups:

1 Exine with a columellar (or predominantly columellar [1]) infratectal structure; polyads distally colporate, if porate, fused distal pores on each grain; pseudocolpi absent; tectum most often undulated, perforated by narrow channels (pits)

1′ Exine with a granular infratectal structure; polyads distally porate; four distal free pores on each grain; tectum thick to very thick, perforated or not

   2 Tectum smooth, flat or undulated; pseudocolpi absent

   2′ Tectum with generally large areoles, either evenly distributed over the distal exine or restricted to local areas; pseudocolpi absent except in a few genera: *Cojoba, Zygia, Marmaroxylon, Obolinga* (Guinet 1989).

These characters allows the distinction of two main groups:

1 The genus *Calliandra* restricted to the species with eight grain polyads (i.e., *Zapoteca* Hernandez 1987 and the Asian-Madagascan species of *Calliandra* excluded)

1′ The other genera of the *Ingeae*

   2 *Albizia* p.p.; *Chloroleucum* p.p.; *Archidendropsis* p.p.; *Enterolobium*

   2′ *Albizia* p.p.; *Abarema*; *Lysiloma, Pithecellobium*; *Havardia*; *Paraserianthes*; *Serianthes*; *Wallaceodendron; Archidendropsis* p.p.; *Pararchidendron*; *Archidendron*; *Zygia, Cojoba*; *Cedrelinga, Klugiodendron, Obolinga*.

Other characters. Apertures in nearly all the *Ingeae* are simple pores; most *Calliandra* s.str. species are colporate. The distribution of the colpori strictly follows the distal symmetry of the grains (four angles: four angular colpori). Further, the colpi are very short and with free ends.

Pseudocolpi (subsidiary colpi) are present in the *Ingeae* only in some species included in the genera *Cojoba, Zygia*, and *Marmaroxylon* (Guinet & Rico 1988) and in the monotypic genus *Obolinga*.

Tectal ornamentation and characteristics. Within the tribe, the tectum is most often even and flat. An undulated tectum is restricted to *Calliandra* s. str. and *Lysiloma* [*L. ambigua* (Vogel) Urb.].

Polyad symmetry in the *Acacieae* and *Ingeae* is under the dependance of the number of associated grains: when this number differ from 4, 12, and 16, the compound units are generally not radially symmetric. Another kind of dissymmetry is the occurrence of heteromorphic central grains. The well-known case has been first described by Erdtman (1952) in *Calliandra* (now: *Zapoteca*) *portoricensis*, where only four central grains in a 16 grain polyad have lens-shaped areas. In *Albizia* (the Neotropical species and the African species of sect. *Zygia*) and in some species of *Lysiloma* (unfortunately the genus has not been revised) the four adjacent central grains on one side of the (16 grain) polyads are strongly areolate while the central grains on the other side are subsmooth or finely areolate.

Pollen characters in the tribe *Mimoseae*

The tribe is taken in the sense of Lewis & Elias (1981), grouping together *Mimoseae* s.str., *Piptadenieae* and *Adenanthereae*. Its pollen diversity is remarkable and includes all the main characters present in the *Acacieae* and *Ingeae*. As only compound grains occur in *Acacieae* and *Ingeae*, comparisons have been limited to the genera of the *Mimoseae* which have this pollen type.

---

[1] Niezgoda & al. (1983) described the presence of few granules in the intercolumellar spaces in *Calliandra emarginata* (H. & B.) Benth.

Main characters. A similar ordering as for the previous tribes leads to the following groupings:
1 Exine with a columellar infratectal structure; compound grains equatorially or distally colporate; pseudocolpi absent; tectum reticulate, smooth or areolate, not undulated: *Fillaeopsis, Calpocalyx, Xylia* p.p., *Schleinitzia, Dichrostachys* p.p., *Gagnebina* p.p., *Leucaena* p.p., *Piptadeniopsis*
1′ Exine without columellae[2]; compound grains equatorially or distally porate; tectum smooth, verrucate or areolate, flat or undulated
    2 Pseudocolpi absent: *Piptadenia, Pseudoprosopis* p.p., *Stryphnodendron* p.p., *Anadenanthera, Gagnebina* p.p., *Adenopodia, Newtonia* sect. *Neonewtonia, Goldmania, Mimosa, Schrankia*
    2′ Pseudocolpi present: *Pseudoprosopis* p.p., *Stryphnodendron* p.p., *Parapiptadenia, Pseudopiptadenia* (*Monoschisma*)

Other characters. Equatorial apertures are a rare feature in the *Mimoseae* with compound grains, e.g., in *Fillaeopsis* (colporate tetrads), *Schleinitzia* (some species), rarely in *Xylia xylocarpa* (sometimes permanent colporate tetrads), but also in *Mimosa* (species with porate tetrads). Pollen characters as in *Mimosa* are frequent in the *Piptadenia* group (as delimited by LEWIS & ELIAS 1981, the genus *Piptadeniopsis* excepted). Non-equatorial colpori are present in *Dichrostachys* (several species), in *Leucaena* only in *L. pluricapitula* (a very isolated species, if included in this genus; cf. WOODSON & SHERRY 1950) and in *Piptadeniopsis*.

Exine. An undulated tectum occurs in the genera included in the *Piptadenia* group, but not in *Piptadeniopsis*.

Well-marked areoles are present in *Calpocalyx, Piptadenia, Pseudopiptadenia, Adenanthera* p.p. and *Mimosa* (a few species, cf. CACCAVARI 1986).

Discussion

There is no difference of opinion about the phyletic position of the *Acacieae* in the *Mimosoideae:* the tribe is always considered a link between *Mimoseae* and *Ingeae*. However, different affinities of the genus *Acacia* taken as a natural unit, have been suggested.

BENTHAM (1875) has viewed *Acacia* and *Lysiloma* (*Ingeae*) as very close and included *Acacia* (*Dugantia*) *rostrata* HUMB. & BONPL. in *Lysiloma* because of its pod characters. NIELSEN (pers. comm.) mentioned that it was difficult to distinguish between *Acacia* sect. *Filicinum* VAS. and some species of *Lysiloma*, except for the pod characters, the stamens in *Lysiloma* often being free nearly to the base.

DNYANSAGAR (1955) has proposed from embryological similarities that *Acacia* s.l., *Pithecellobium* s.l. and *Calliandra* s.l. belong to the same alliance, with *Acacia* as the basic genus. D'ARCY (1971) has suggested from the study of *Acacia anegadensis* a possible relationship with *Pithecellobium*, and GUNN (1984) has emphasized that in the *Mimosoideae* only *Acacia* and *Pithecellobium* have seeds which may have arils. ROBBERTSEE (1974) has compared the sometimes apocarpous *Acacia nigrescens* OLIV. and *A. mellifera* (VAHL) BENTH., two African species belonging to the subgen. *Aculeiferum*, with *Archidendron* (*Ingeae*). PEDLEY (1986) supported

---

[2] More TEM data are needed in this group. However, careful observations in LM shows that columellae are absent and, when TEM data are available, the exine structure is granular.

affinities of *Senegalia* and *Racosperma* (*Acacia*) with *Calliandra* s.l. By contrast, Ross (1979) has mentioned relationships between *Acacia* s.l. and *Prosopis* (*Mimoseae*). SORSA (1969) and GUINET (1969) have observed close pollen similarities between *Acacia* s.l. and *Piptadeniopsis* (*Mimoseae*).

Acacia compared to Lysiloma. If these genera are viewed as the closest, *Lysiloma* would be related only with the subgen. *Aculeiferum* of *Acacia*. In both groups exine structure and apertures are of the same type. However, polyad dissymmetry puts *Lysiloma* nearer to *Albizia*.

Acacia compared to Pithecellobium. As both *Acacia* subgen. *Acacia* and *Pithecellobium* s.str. have spinescent stipules, one could suspect affinities between these two taxa. However, their pollen characters do not suggest relationships. *Acacia* subgen. *Acacia* has distally tricolporate pollen with columellar exine structure. *A. (Fishlockia) anegadensis* BRITT. also has these characters which does not argue for its transfer to the *Ingeae* (cf. D'ARCY 1971, PEDLEY 1986). *Pithecellobium* pollen is devoid of colpori, has a granular exine structure and differs from *Acacia* by its areolate exine surface.

Acacia compared to Archidendron. The suggestion (ROBBERTSEE 1974) that these genera must be related on the grounds that pluricarpellate condition may occur in both, is ruled out, both from pollen morphology and general characters (cf. NIELSEN & al. 1984). However, the genus *Archidendropsis* (undescribed at the time of ROBBERTSEE's work) has three species: *A. basaltica* (MUELL.) NIELS., *A. thozetiana* (MUELL.) NIELS. and *A. xanthoxylon* (WHITE & FRANCIS) NIELS., all belonging to the subgen. *Basaltica* NIELS. (Australia, New Caledonia, New Guinea), formerly included either in *Albizia* or in *Acacia*. They have (caducous) stipules, often developed into small stipular spines (NIELSEN & al. 1983). The pollen is a small 16 grain polyad with a granular exine structure and porate apertures, a smooth perforated tectum, as in several species of *Albizia* and *Acacia* subgen. *Aculeiferum*.

Acacia compared to Calliandra. When *Acacia* and *Calliandra* s.str. are viewed as related genera, the greater pollen similarity is between *Calliandra* and *Acacia* subgen. *Acacia*: they share a columellar exine structure, infrequent in their tribes. When Ross (1974) and ROBBERTSEE & VON TEICHMAN (1979) described the morphology of *Acacia redacta* Ross (now: *Calliandra redacta*, cf. THULIN & al. 1981), they emphasized that this species was unusual in *Acacia* (stamens united in a basal tube), but had some characters in common with *Acacia* (*Faidherbia*) *albida*, i.e., stipular spines. As stated above, *Faidherbia albida* cannot be included in *Acacia* subgen. *Acacia*. The pollen differences between *Acacia* s.l. and *Calliandra* s.str. are so numerous that they do not suggest close relationship between the two genera. NEVLING (in GOLDBLATT 1981) has suggested affinities of the genus *Calliandra* with the *Mimoseae*, perhaps with *Xylia*. However, the floral differences are very great and the stamen content in *Calliandra* is the same as in *Acacieae* and *Ingeae* (8 polyads per stamen), while in *Xylia*, 30 – 50 polyads per stamen are present, as in many *Mimoseae* genera with compound grains.

Acacia compared to Prosopis. The genera *Acacia* and *Prosopis* (cf. Ross 1979) appear only distantly related both from their floral characters (ten free stamens) and their pollen characters (single, equatorially tricolporate grains). BURKART (1976) has demonstrated a remarkable parallelism between the two genera, but this is restricted to the nature of stipules and the presence/absence of prickles.

Acacia compared to Piptadeniopsis. The monotypic genus *Piptadeniopsis* (spinescent stipules, flowers in heads, no involucel) has been referred either to the *Prosopis* group (LEWIS & ELIAS 1981) or to the *Piptadenia* group (BURKART 1944, GUINET 1981). Pollen similarities between this genus and the subgen. *Acacia* of *Acacia* are striking. Both have a columellar exine structure and three distally fused colpori. The exine is areolate, as in several genera of the *Piptadenia* group and in most *Ingeae*, but unlike *Acacia*, and the arrangement of the grains within the polyads is typical of what occurs in many genera of the *Mimoseae*. Here again, the floral characters (ten free stamens) and the stamen content (50 – 60 polyads per stamen in *Piptadeniopsis*) makes the difference.

Conclusion

PEDLEY (1986) in a comprehensive survey of the best known characters in the genus *Acacia* s.l. has suggested an independant derivation of two groups: *Acacia* (*A.* subgen. *Acacia*) in one, *Senegalia* and *Racosperma* (*Acacia*, subgg. *Aculeiferum* and *Phyllodineae*) in the other. The two groups are viewed as derived from different lines of the *Ingeae*.

Individualization of two main groups is supported by pollen morphology (GUI-NET 1981) and clearly isolates *Acacia* (*A.* subgen. *Acacia*). The absence of transitions between these groups reflects the lack of close relationship. When emphasis is given to the pollen characters, the generic distinction between *Senegalia* and *Racosperma* is not obvious: ± 100 Australian *Racosperma* spp. with phyllodes but with the main pollen characteristics as found in *Senegalia* weakens the discrimination. In spite of numerous similarities of *Acacia* s.l. with the *Ingeae*, pollen characters suggest that the genus (or genera) must instead be viewed as an early offshoot of the tribe *Mimoseae*. As emphasized above, practically all of the pollen characters found in *Acacieae* and *Ingeae* are present in the *Mimoseae*. However, it is known (GUINET & FERGUSON 1987) that in the whole *Leguminosae* family the change in exine structure (from columellar to granular) is frequently associated with the change in the nature of the aperture (from colpori to pori), and such changes have occurred in a repetitive manner in many genera: the same model is also present in *Acacia* s.l. However, (1) in the *Leguminosae*, a family where the link between exine structure and apertural type is best documented, the change of columellar-colporate to granular-porate does not affect aperture spacing. In the tricolporate-columellar *Mimosoideae*, the apertures are regularly distributed, and equidistant most generally in simple pollen but also (less frequently) in permanent polyads. Such a symmetry lacks in *Acacieae* and *Ingeae* but is present in *Acacia* subgen. *Acacia*. Its occurrence here cannot be explained without reference to the *Mimoseae* tribe, especially to *Piptadeniopsis*. (2) Granular-porate pollen grains, which consistently occur in *Acacia* subgg. *Aculeiferum* and *Phyllodineae*, are present in all of the *Ingeae, Calliandra* s.str. excepted. In this later genus, when colpori are present they are located at the distal angles of the grains and distributed at 90° angles, as are the distal angular pores in the other genera of the same tribe. Furthermore, the pollen characters of *Calliandra* are very distinct as compared to *Acacia* (calymmate highly dissymmetric polyads). However, granular-porate pollen grains are also present in the *Mimoseae*. In *Acacia*, the small group *Filicinae* BENTH. (granular-porate grains) has very poorly differentiated pollen characters and an undulated tectum, as is most frequent in the *Piptadenia* group of the *Mimoseae*. The *Filicinae* are also distinct from *Acacia* by the absence of foliar glands, prickles and stipular spines and by their biochemical characters (free amino-acids of seeds; cf. EVANS in PEDLEY 1986). RICO (1987) has listed "stipels" in some species and in the genus *Mimosa* BARNEBY (1985) has described "paraphyllidia", a name for the rudimentary leaflets formerly called stipels by BENTHAM. This distinct *Acacia* group appears the nearest to the *Mimoseae* tribe, from which it remains separated by the low stamen content (anthers containing eight polyads). Thus, the tribe *Acacieae* is shown to be a grade rather than a clade (KANIS 1986), and the *Piptadenia* group of the *Mimoseae* appears basic for understanding the pollen characters in *Acacia* s.l., even if the stamen content provides one of the best characters to separate the *Mimoseae* from the *Acacieae* and *Ingeae,* irrespective of stamen fusion and number. Accepting an origin of *Acacia* s.l. among different lines of the *Piptadenia* group would better explain the long disputed phyletic significance of many characters in *Acacia* when viewed as a natural unit, with *Acacia* subgen. *Acacia* interpreted either as basic or as derived (cf. Ross 1981). The *Mimoseae* tribe represents the polymorphic and older core of the subfamily and has a significantly higher pollen diversity, sometimes even within a single genus. In the most advanced tribes *Acacieae* and *Ingeae* the loss in pollen diversity renders the generic pollen characters more stable and thus increases their value and use for taxonomic purposes.

References

BARNEBY, R. C., 1985: The genus *Mimosa* (*Mimosaceae*) in Bahia, Brazil: new taxa and nomenclatural adjustments. — Brittonia **37**: 125–153.

BELL, E. A., EVANS, C. S., 1978: Biochemical evidence of a former link between Australia and the Mascarene Islands. — Nature **273**: 295–296.

BENTHAM, G., 1875: Revision of the suborder *Mimoseae.* — Trans. Linn. Soc. London **30**: 335–650.

BRITTON, N. L., ROSE, J. N., 1928: North American Flora **23**: 1–194. — New York: New York Botanical Gardens.

BURKART, A., 1944: Tres nuevas leguminosas del Paraguay. — Darwiniana **6**: 477–493.

— 1976: A monograph of the genus *Prosopis* (*Leguminosae* subfam. *Mimosoideae*). — J. Arnold Arbor. **57**: 219–249.

CACCAVARI, M. A., 1986: Estudio de los caracteres del polen en las *Mimosa-Lepidotae.* — Pollen & Spores **28**: 29–42.

D'ARCY, W. G., 1971: The island of Anegada and its flora. — Smithsonian Institution, Atoll Research Bull. **139**: 1–21.

DNYANSAGAR, V. R., 1955: Embryological studies in the *Leguminosae,* 11: embryological features and formula and taxonomy of the *Mimosaceae.* — J. Indian Bot. Soc. **34**: 362–374.

ERDTMAN, G., 1952: Pollen morphology and plant taxonomy. Angiosperms. — Stockholm: Almqvist & Wiksell.

GOLDBLATT, P., 1981: Cytology and the phylogeny of *Leguminosae.* — In POLHILL, R. M., RAVEN, P. H., (Eds.): Advances in legume systematics **2**, pp. 427–463. — Kew: Royal Botanic Gardens.

GUINET, PH., 1964: Données nouvelles sur le rôle de la morphologie du pollen dans la classification du genre *Acacia.* — Compt. Rend. Acad. Sci. **258**: 4823–4824.

— 1969: Les Mimosacées, étude de palynologie fondamentale, corrélations, évolution. — Inst. Franç. Pondichéry, Trav. Sect. Sci. Techn. **9**: 1–293.

— 1979: Biologie évolutive dans le genre *Acacia*: arguments palynologiques. — Ecole Pratique des Hautes Etudes, Mémoires et Travaux de l'Institut de Montpellier **4**: 41–51.

- 1981: *Mimosoideae:* the characters of their pollen grains. – In POLHILL, R. M., RAVEN, P. H., (Eds.): Advances in legume systematics **2**, pp. 835–857. – Kew: Royal Botanic Gardens.

- 1986: Geographic patterns of the main pollen characters in genus *Acacia* (*Leguminosae*), with particular reference to subgenus *Phyllodineae*. – In BLACKMORE, S., FERGUSON, I. K., (Eds.): Pollen and spores: form and function. – Linn. Soc. Symp. Ser. **12**: 297–311.

- 1989: Pollen of *Obolinga zanonii* (*Mimosaceae*) – Brittonia **41**: 173–174.

- LUGARDON, B., 1976: Diversité des couches de l'exine dans le genre *Acacia* (*Mimosaceae*). – Pollen & Spores **18**: 483–511.

- FERGUSON, I. K., 1987: Structure, evolution and biology of pollen in *Leguminosae*. – Ann. Missouri Bot. Gard. (in press).

- RICO, L., 1988: Pollen characters in the genera *Zygia*, *Marmaroxylon* and *Cojoba* (*Leguminosae, Mimosoideae, Ingeae*): a comparison with related genera. – Pollen & Spores **30**: 313–328.

GUNN, C. R., 1984: Fruits and seeds of genera in the subfamily *Mimosoideae* (*Fabaceae*). – U.S. Department of Agriculture, Agricultural Research Service, Tech. Bull. **1681**: 1–194.

HERNANDEZ, H. L., 1987: *Zapoteca:* a new genus of neotropical *Mimosoideae*. – Ann. Missouri Bot. Gard. **73**: 755–763.

KANIS, A., 1986: Taxonomic changes in Australian *Mimosaceae*. – Austral. Syst. Bot. Soc. Newsletter **48**: 9–13.

LEWIS, G. P., ELIAS, T. S., 1981: *Mimoseae* BRONN. – In POLHILL, R. M., RAVEN, P. H., (Eds.): Advances in legume systematics **2**, pp. 155–168. – Kew: Royal Botanic Gardens.

MAIRE, R., 1987: Flore de l'Afrique du Nord, 16: Leguminosae. pp. 1–302. – Paris: Lechevalier.

NIELSEN, I., 1981: *Ingeae* BENTH. – In POLHILL, R. M., RAVEN, P. H., (Eds.): Advances in legume systematics **2**, pp. 173–190. – Kew: Royal Botanic Gardens.

- GUINET, PH., BARETTA-KUIPERS, T., 1983: Studies in the Malesian, Australian, and Pacific *Ingeae* (*Leguminosae-Mimosoideae:* the genera *Archidendropsis, Wallaceodendron, Paraserianthes, Parachidendron* and *Serianthes*). – Bull. Mus. Natn. Hist. Nat. Paris, 4e Ser., sect. B, Adansonia **5**: 335–360.

- – 1984: Studies in the Malesian, Australian and Pacific *Ingeae* (*Leguminosae* – *Mimosoideae:* the genera *Archidendropsis, Wallaceodendron, Paraserianthes, Parachidendron,* and *Serianthes*). – Bull. Mus. Natn. Hist. Nat. Paris, 4e Ser., sect. B, Adansonia **6**: 79–111.

NIEZGODA, C. J., FEUER, S. M., NEVLING, L. I., 1983: Pollen ultrastructure of the tribe *Ingeae* (*Mimosoideae: Leguminosae*). – Amer. J. Bot. **70**: 650–667.

PEDLEY, L., 1975: Revision of the extra-Australian species of *Acacia* subgen. *Heterophyllum*. – Contrib. Queensland Herbarium **18**: 1–24.

- 1978: A revision of *Acacia* MILL. in Queensland. – Austrobaileya **1**: 175–234.

- 1986: Derivation and dispersal of *Acacia* (*Leguminosae*), with particular reference to Australia, and the recognition of *Senegalia* and *Racosperma*. – Bot. J. Linn. Soc. **92**: 219–254.

RICO, L., 1987: *Acacia sousae* (*Leguminosae: Mimosoideae*), a new species from Mexico. – Brittonia **39**: 130–132.

ROBBERTSEE, P. J., 1974: The genus *Acacia* in South Africa II. With special reference to the morphology of the flower and inflorescence. – Phytomorphology **24**: 1–15.

- VON TEICHMAN, I., 1979: The morphology of *Acacia redacta* J. H. Ross. – J. South African Bot. **45**: 11–23.

Ross, J. H., 1974: Notes on *Acacia* species in Southern Africa: 4. – Bothalia **11**: 231–234.
– 1979: A conspectus of the African *Acacia* species. – Mem. Bot. Survey South Africa **44**: 1–155.
– 1981: An analysis of the African *Acacia* species: their distribution, possible origins and relationships. – Bothalia **13**: 389–413.
Sorsa, P., 1969: Pollen morphological studies on the *Mimosaceae*. – Ann. Bot. Fenn. **6**: 1–34.
Thulin, M., Guinet, Ph., Hunde, A., 1981: *Calliandra* (*Leguminosae*) in continental Africa. – Nordic J. Bot. **1**: 27–34.
Vassal, J., 1972: Apport des recherches ontogéniques et séminologiques à l'étude morphologique, taxonomique et phylogénique du genre *Acacia*. – Bull. Soc. Hist. Nat. Toulouse **108**: 125–247.
Woodson, R. E., Schery, R. W., 1950: Flora of Panama 5-2. – Ann. Missouri Bot. Gard. **37**: 121–314.

Address of the author: P. Guinet, Laboratoire de Palynologie E.P.H.E., Université des Sciences de Techniques du Languedoc, F-34095 Montpellier, France.

Pl. Syst. Evol. [Suppl. 5], 91 – 102 (1990)

Taxonomic and evolutionary significance of pollen morphology in the *Apocynaceae*

S. NILSSON

Received January 26, 1988

Key words: Angiosperms, *Apocynaceae*. – Pollen, ultrastructure, taxonomy, evolution.

Abstract: The pollen morphology of 17 genera from nine tribes of the *Apocynaceae* was investigated with special reference to taxonomy. In the *Plumeroideae*, the *Ambelanieae* and the *Macoubeeae* are alike in pollen morphology, while the *Tabernaemontaneae* differ. Also, the *Catharanthinae* genera of *Plumerieae* deviate except *Amsonia* and *Rhazya*. *Holarrhena* (*Plumerieae*) resembles *Alafia, Pleioceras*, etc. (*Apocynoideae*) to which subfamily it should possibly be transferred. *Allamanda* (*Plumeroideae: Allamandeae*) is similar to *Cerberoideae* (particularly *Thevetia*). From their pollen morphology the *Apocynoideae* appear more advanced than the *Plumeroideae*.

The *Apocynaceae* can be subdivided into three subfamilies, *Plumerioideae, Cerberoideae* and *Apocynoideae*, with 14 tribes and numerous subtribes (Table 1; LEEUWENBERG oral. comm.). In the present study pollen of nine tribes was examined with special reference to taxonomy and evolutionary aspects. The significance of pollen morphology in the *Apocynaceae* was discussed in previous papers, e.g., NILSSON & VAN CAMPO (1978), VAN CAMPO & al. (1979), and NILSSON (1986).

Material and methods

Pollen material was obtained from the herbaria BR, K, L, MO, P, S, US, and WAG.

Specimens examined. *Alafia lucida* STAPF. Gabon. LE TESTU 7953 (P). *Allamanda cathartica* L. Gabon. LEEUWENBERG 12540 (WAG). *A. puberula* A. DC. var. *glabrata* MUELL. ARG. Brazil. HARLEY & al. 16385 (US). *Ambelania acida* AUBL. Brazil. EGLER & IRWIN 46545 (US). *Amsonia ciliata* WALT. U.S.A., Georgia, SASSEN s.n. (WAG). *Catharanthus roseus* (L.) G. DON. Liberia. VAN HARTEN 29 (WAG). *Cerbera floribunda* K. SCHUM. New Guinea. SCHODDE 4536 (L); NGF 24770 (L). *C. manghas* L. Taiwan. HUANG 9156 (S). *Holarrhena floribunda* (G. DON) DUR. and SCHINZ. Ghana. LEEUWENBERG 11046 (WAG). *H. pubescens* (BUCH.-HAM.) WALL. ex G. DON. Burma. CHIN 6830 (MO). *Kopsia flavida* BLUME. Indonesia. No collector. (US). *Macoubea guianensis* AUBL. Brazil. DUCKE 1006 (K); PIRES 5524 (MO). *M. sprucei* (MUELL. ARG.) MARKGR. var. *pauciflora* (SPRUCE ex MUELL. ARG.) MONACH. Colombia. SASTRE 2340 (P). *Molongum grandiflorum* (HUBER) PICHON. Brazil. PIRES & BLACK 1471 (P). *M. laxum* (BENTH.) PICHON. Colombia. MAGUIRE, WURDACK & BUNTING 36267 (S). *Ochrosia mariannensis* A. DC. Guam. D. ANDERSON 199 (US). *Pleioceras barteri* BAILL. Ivory Coast. LEEUWENBERG 12152 (WAG). *Rhazya stricta*

Table 1. Provisional classification of *Apocynaceae* after Leeuwenberg (1989, oral. comm.). In bold type, tribes investigated

Plumerioideae	*Cerberoideae*	*Apocynoideae*
Carisseae	**Skytantheae:** *Skytanthus*	Echitheae
Chilocarpeae	**Cerbereae:** *Cerbera,*	**Nerieae:** *Pleioceras, Alafia*
Ambelanieae: *Ambelania, Molongum*	*Thevetia*	Apocyneae
Macoubeeae: *Macoubea*		Ichnocarpeae
Tabernaemontaneae: *Tabernaemontana*		
Plumerieae: *Rhazya, Amsonia, Catharanthus, Vinca, Holarrhena*		
Rauvolfieae: *Ochrosia, Kopsia*		
Allamandeae: *Allamanda*		

Decne. Saudi Arabia. Schimper 812 (L). *Skytanthus hancorniaefolius* (A. DC.). Benth. & Hook. Brazil. Magalhaes 18312 (US). *S. martianus* (Muell. Arg.) Miers. Brazil. Pickel s.n. (US). *Tabernaemontana undulata* Vahl. Surinam. Boswezen 5470 (U). *T. ventricosa* Hochst. ex A. DC. Tanzania. Semsei 1947 (BR). *Thevetia ahouai* A. DC. Guatemala. Pither 256 (US). *T. amazonica* Ducke. Bolivia. Cardenas 3755 (US). *T. peruviana* (Pers.) K. Schum. Peru. Landeman 4943 (K). *Vinca minor* L. Sweden, Sollentuna 1985. S. Nilsson s.n. (cult.).

Methods. For light microscopy (LM) the pollen samples were acetolysed, mounted in glycerine jelly and sealed with paraffine. For SEM, entire and fractured pollen grains were

Fig. 1. *a Tabernaemontana undulata*. Pollen grain in equatorial view showing a short, narrow colpus, an equatorial oral band bordered by thickened nexine (costae), and filled with granules. Nexine densely granular. LM, × 1 000. *b* The same in optical cross-section. LM, × 1 000. *c T. ventricosa*. Pollen grain in equatorial view showing a relatively short colpus with acute ends and a well delimited lalongate, oval-shaped os. LM, × 1 000. *d* The same in optical cross-section. LM, × 1 000. *e Molongum laxum*. Aspidote pollen grain in equatorial view with extremely short colpi and ora with thickened margin. LM, × 1 000. *f M. laxum*. Pollen grain in polar view, angular in outline. LM, × 1 000. *g Macoubea guianensis* (Pires & al. 5524; MO). Aspidote, 2-colporate pollen grain in equatorial view, inner contour dumb-bell-shaped in outline. LM, × 1 000. *h Ambelania acida*. Aspidote pollen grain in polar view with protruding aperture margins, and smooth, densely perforated sexine. SEM, × 1 500. *i Molongum grandiflorum*. Aspidote pollen grain in equatorial view showing short colpi with acute ends, ora, and smooth, distinctly and densely spaced, perforated sexine. SEM, × 1 100. *j Macoubea guianensis* (Ducke 1006; K). Aspidote pollen grain in slightly oblique polar view with smooth, densely perforated sexine. SEM, × 1 500. *k M. guianensis* (Ducke 1006; K). A short, wide colpus with rounded ends and granular membrane, delimited by protruding sexine margin. Os relatively large. Sexine smooth, densely perforated except at the colpus margin. SEM, × 3 600

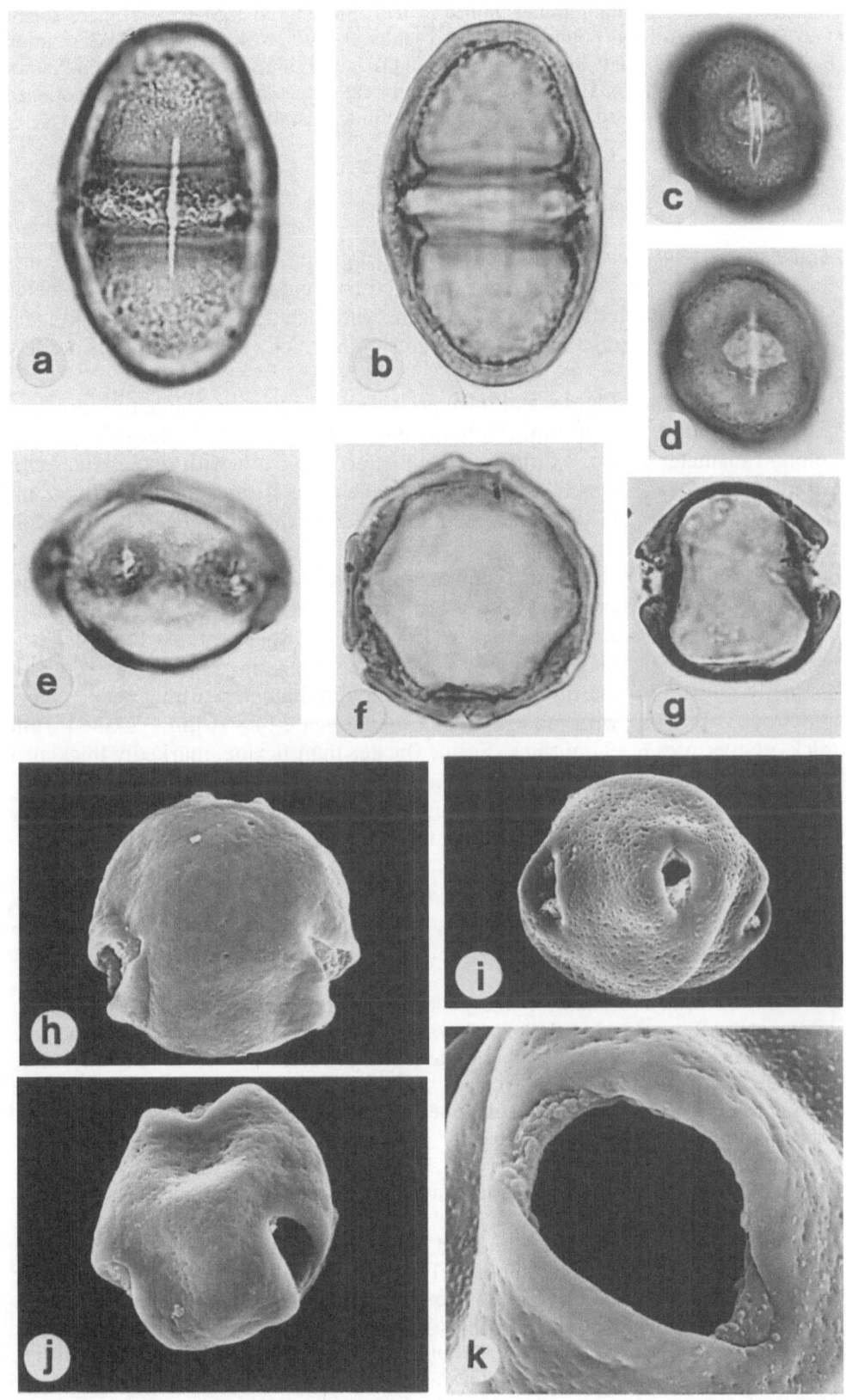

coated with gold-palladium and examined in Jeol JSM 35 and JSM-25 S 11 microscopes. For TEM, unacetolysed polliniferous dried material was rehydrated in a TAG solution buffered in Na-Cacodylate buffer, post-fixed in OsO_4 and embedded in epoxy resin (Spurr 1969), or Epon-Araldite. Ultrathin sections were cut with a diamond knife, post-stained in uranyl acetate and lead citrate, and examined in a Zeiss 10 A microscope.

Pollen descriptions

Plumeroideae

Ambelanieae. *Ambelania acida* (Fig. 1 h). Pollen grains 3-colporate, suboblate, distinctly aspidote, $25 \times 30 \,\mu$m. Amb rounded-triangular. Apocolpium diameter $18 \,\mu$m. – Colpi relatively short, $9 \times 5 \,\mu$m with acute ends. Ora lalongate $(4 \times 8 \,\mu$m). – Exine 3 μm thick, at apertures 5 μm thick. Sexine smooth, densely perforate.

Molongum laxum (Fig. 1 e, f). Pollen grains 3 – 5-colporate, oblate to suboblate, distinctly aspidote, $33 \times 44 \,\mu$m. Amb angular (pentangular to hexangular). Apocolpium diameter 25 μm. – Colpi relatively short, $8 \times 3 \,\mu$m⸱ with acute ends. Ora lalongate $(3 \times 5 \,\mu$m), encircled by a $2 – 3 \,\mu$m wide annular thickening. – Exine 2 μm thick, at apertures 4 μm thick. Sexine thicker than nexine, markedly, thickened at the apertures, smooth, perforate.

M. grandiflorum (Fig. 1 i) has similar pollen grains with smooth, densely and distinctly perforate sexine; perforations of unequal size.

Macoubeeae. *Macoubea guianensis* (Fig. 1 g, j, k). Pollen grains mostly 2-colporate, oblate to suboblate, distinctly aspidote, $25 \times 33 \,\mu$m. Amb oval-shaped with distinctly protruding aperture margin. Apocolpium diameter 16 μm. – Colpi short and broad, $10 \times 6 \,\mu$m, with rounded ends. Ora lalongate $14 \times 16 \,\mu$m. – Exine 1.5 μm thick, at apertures $6 – 7 \,\mu$m thick. Sexine thicker than nexine, markedly thickened at apertures, smooth, densely perforate.

M. sprucei var. *pauciflora*. Pollen grains usually 3-colporate, suboblate, distinctly aspidote, $25 \times 27 \,\mu$m. Amb rounded-triangular. Apocolpium diameter $16 – 18 \,\mu$m. Otherwise similar to *M. guianensis*.

Tabernaemontanae. *Tabernaemontana undulata* (Fig. 1 a, b; Van Campo & al. 1979). Pollen grains 3-colporate, prolate, $60 \times 35 \,\mu$m. Amb triangular. – Colpi rel-

---

Fig. 2. *a Holarrhena floribunda*. 3-porate pollen grain in oblique polar view. Sexine smooth, indistinctly perforate. SEM, × 2 000. *b* The same, showing a pore with granules at the pore margin. Sexine smooth, indistinctly perforate. SEM, × 6 400. *c Pleioceras barteri*. 3-porate pollen grain in polar view. Sexine smooth. SEM, × 1 700. *d* The same, showing a pore surrounded by smooth, sparsely perforate sexine. SEM, × 6 000. *e Holarrhena pubescens*. Exine consisting of a distal stratum ("tectum") subtended by irregular, partly fused granular elements delimited by osmiophilic substance. Intine indistinctly structured. TEM, × 12 500. *f Alafia lucida*. Exine consisting of a distal stratum ("tectum") to which are adhered granular elements of unequal size and shape. Intine with numerous tubules in the outer part. TEM, × 12 500. *g Amsonia ciliata*. Exine stratified into tectum, a granular-reticulate infratectal stratum, subtended by a sole. Intine fibrillar. TEM, × 10 000. *h Vinca minor*. Exine extremely thin consisting of a single or possibly two-layered coat subtended by markedly thick, stratified fibrillar intine. TEM, × 12 500. *i Catharanthus roseus*. Exine stratified into a discontinuous tectum, a granular-reticulate infratectal stratum; mescolpial plates consisting of a thin sole subtended by endexine. TEM, × 6 700

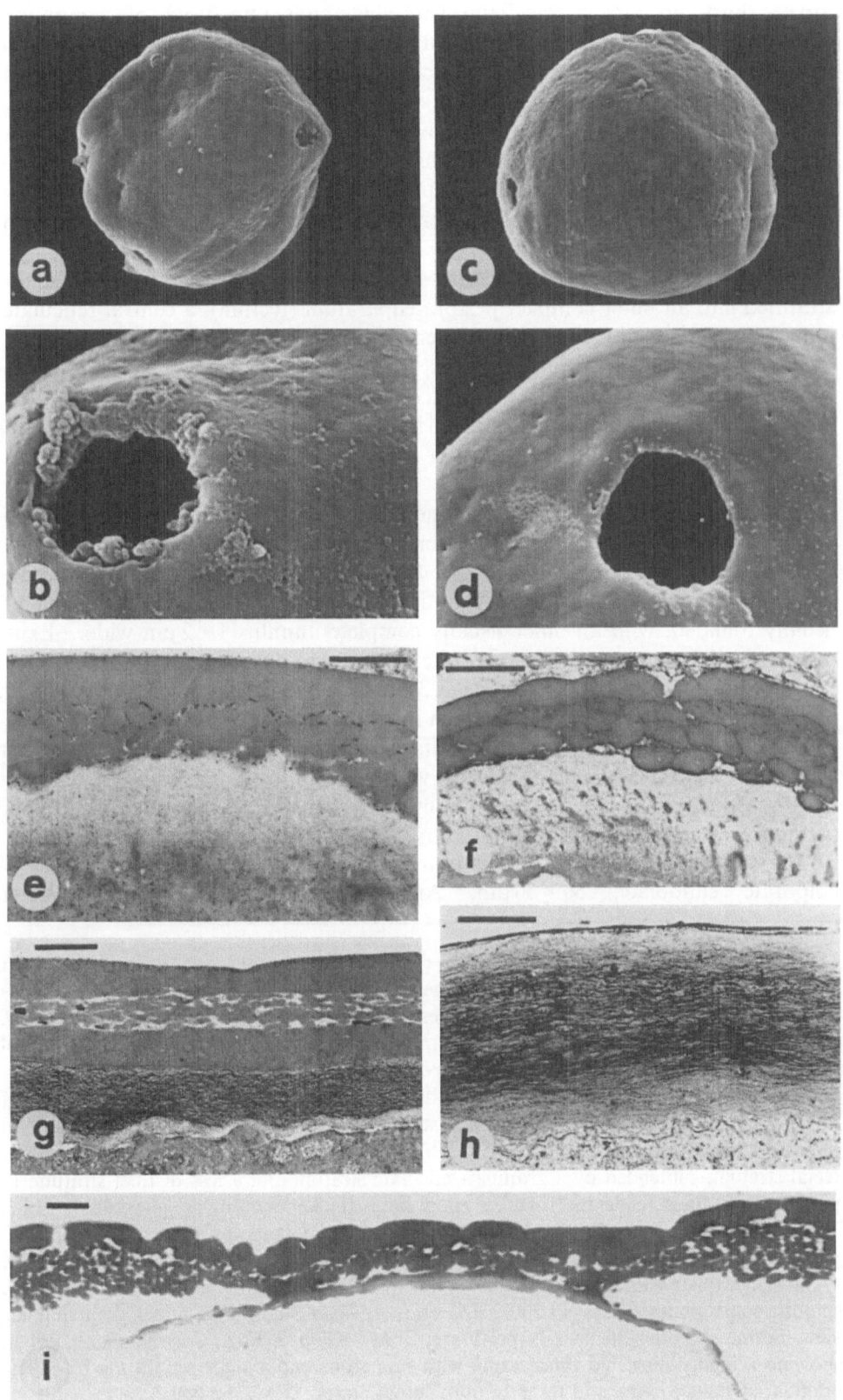

atively short and narrow, 25 × 2 µm, with acute ends. Ora fused to a continuous equatorial band, delimited by thickened nexine margins and filled with numerous granules. – Exine 2 µm thick. Sexine thicker than nexine, smooth, perforate. Inside of exine covered by grains of different sizes.

T. ventricosa (Fig. 1 c, d; VAN CAMPO & al. 1979). Pollen grains 3-colporate, prolate spheroidal, 30 × 28 µm. Amb rounded-triangular. – Colpi short, 15 × 2 µm, with acute ends. Ora lalongate (6 × 11 µm), surrounded by annular nexine thickening. – Exine 3 µm thick. Sexine thicker than nexine, smooth, perforate. Inside of exine granular.

Plumerieae: *Catharanthinae. Amsonia ciliata* (Fig. 2 g; NILSSON 1986). Exine stratified into an outer compact perforated stratum (tectum), a central reticulate-granular stratum, subtended by an inner compact stratum (sole). Intine fibrillar.

Rhazya stricta has similar ultrastructure.

Catharanthus roseus (Fig. 2 i; NILSSON 1986). Exine stratified into an outer, densely perforated stratum (tectum), and inner granular-reticulate stratum, in places subtended by a thin stratum of ectexine (sole) and endexine forming mesocolpial plates.

Vinca minor (Fig. 2 h). Exine extremely thin (c. 0.5 µm). Intine multi-layered, dominated by two variously stained, fibrillar strata.

Holarrheninae. Holarrhena floribunda (Fig. 2 a, b). Pollen grains 3-porate, spheroidal to irregularly shaped, 30 µm in diameter. – Pores 2 – 4 µm in diameter, usually rounded, with an inner usually complete annulus 1 – 2 µm wide. – Exine c. 1 µm thick, smooth, sparsely perforate, with numerous granules on the inside.

H. pubescens (Fig. 2 e). Pollen grains 3-porate, spheroidal to irregularly shaped, 28 – 35 µm in diameter. – Pores 2 – 4 µm in diameter, usually rounded, with a narrow annulus. – Exine 1 µm thick, smooth, sparsely perforate, consisting of an outer stratum with irregular inner surface to which are adpressed granules of different size and shape, partly fused to larger complexes. Between the granules osmiophilic substance. – Intine with an outer part and tubules, and an inner fibrillar part.

Rauvolfieae: *Ochrosiinae. Ochrosia mariannensis* (Fig. 3 a). Pollen grains 3-colporate suboblate, 50 × 60 µm. Amb rounded. Apocolpium diameter

Fig. 3. *a Ochrosia mariannensis.* 3-colporate, verrucose pollen grain in slightly oblique, polar view. The verrucae are comparatively large, partly fused and aligned along the colpus margins and at the poles, smaller in the mescolpial areas. SEM, × 800. *b Kopsia flavida.* Exine homogeneous, subdivided into two strata separated by a discontinuous commissural line of osmiophilic substance. Intine stratified, fibrillar. TEM, × 14 800. *c Skytanthus martianus.* 3-colporate pollen grain in slightly oblique polar view. Colpus membrane granular, divided by a longitudinal fissure, os very distinct. Sexine smooth, perforated, depressed in the centre of mesocolpia. SEM, × 1 400. *d S. hancorniaefolius.* Exine consisting of a thin, tectal stratum, subtended by a granular-reticulate stratum and a less distinct stratum, in part darkly stained (endexine?). Intine many times thicker than exine, structured. TEM, × 12 500. *e Cerbera floribunda.* 3-colporate pollen grain in polar view. Sexine uneven, perforate, except at the colpus margins, fused to smooth areas at the poles. SEM, × 700. *f* The same. Polar fragment of a pollen showing densely granular inner exine and three smooth, colpal plates. SEM, × 1 300. *g Thevetia peruviana.* 3-colporate pollen grain in polar view. Sexine smooth and densely perforate. SEM, × 700. *h* Polar fragment of a pollen showing a highly dissected inner exine with numerous endocracks separating irregular, granular-spinulose islets, and three smooth colpal plates. SEM, × 1 000

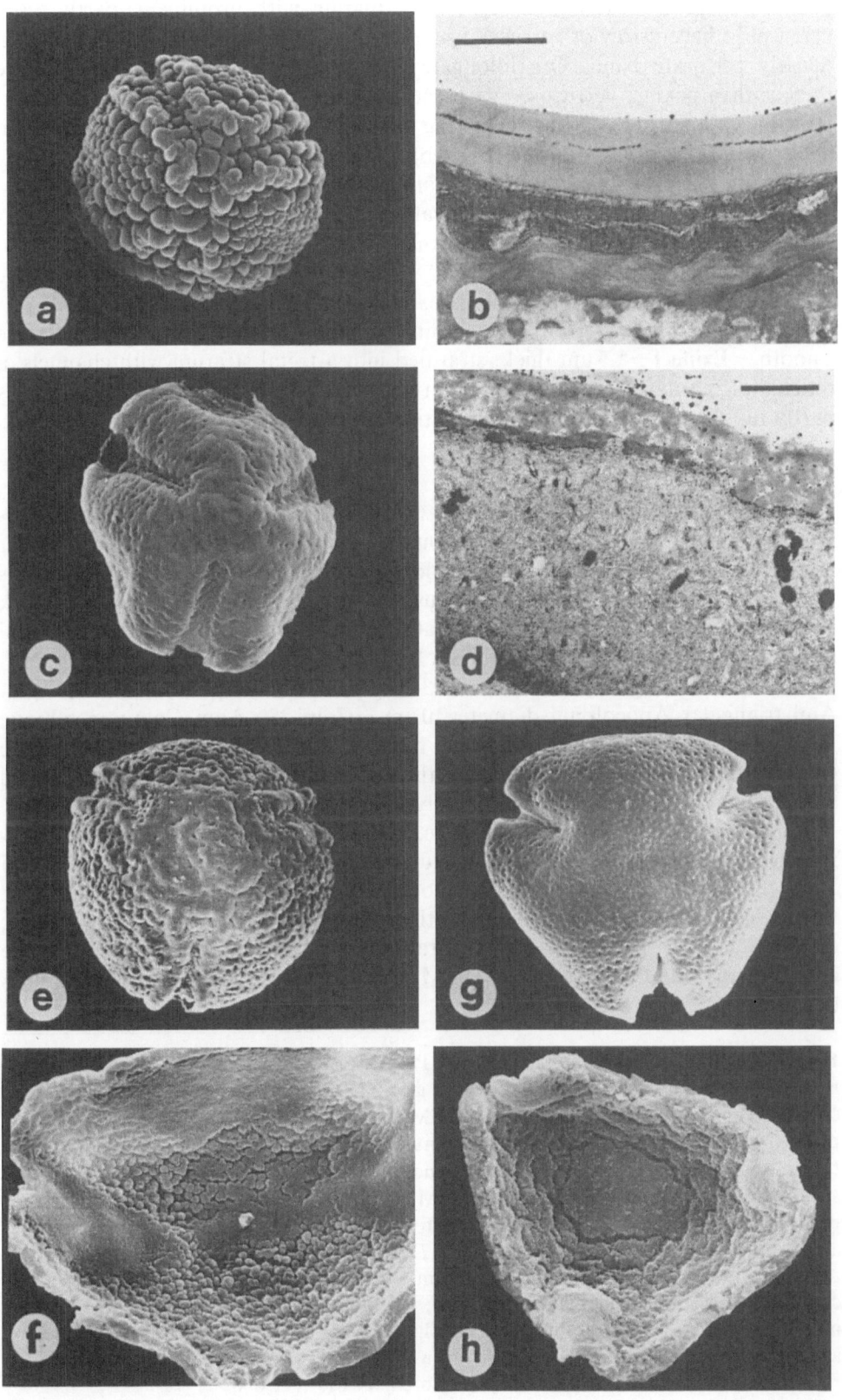

35 µm. – Colpi 20 × 3 µm with blunt ends, margin with prominent, partly fused verrucae; adjacent row of verrucae usually completely fused forming a thickened, densely perforate band. Ora lolongate (10 × 3 µm). – Exine 4 µm thick. Sexine thicker than nexine, verrucose. Verrucae 2 – 3 µm high, 2 – 8 µm wide, relatively large at the aperture margin, decreasing in size towards the centre of mesocolpia.

Vallesiinae. Kopsia flavida (Fig. 3 b). Exine c. 1 µm thick, homogeneous, subdivided by a darkly stained commissural line. Intine stratified, fibrillar.

Allamandeae. *Allamanda puberula* var. *glabrata* (Fig. 4 a). Pollen grains 3-colporate, oblate spheroidal, 72 × 77 µm. Amb rounded-triangular to triangular. Apocolpium diameter 28 – 30 µm. – Colpi 40 × 2 µm. Ora lolongate (16 × 3 µm). – Exine 1.5 – 2 µm thick. Sexine thicker than nexine, smooth, densely perforate.

A. cathartica (Fig. 4 b). Pollen grains 3-colporate (much wrinkled). Sexine smooth. – Exine 1 – 1.5 µm thick, stratified into a tectal stratum with channels, a reticulate-granular infratectal stratum, at the base of which there are larger granules partly fused to a discontinuous sole. Intine structured, fibrillar.

Cerberoideae

Skytantheae. *Skytanthus hancorniaefolius* (Fig. 3 d). Pollen grains 3-colporate, prolate spheroidal, 43 × 38 µm. Amb rounded-triangular. Apocolpium diameter 12 µm. – Colpi 33 × 2 µm. Ora usually lolongate (5 × 3 µm) or lalongate (not always traceable due to equatorial sexine protrusions). – Exine 3 µm thick. Sexine as thick as nexine, scabrate; stratified into a tectal stratum, a granular-reticulate infratectal stratum, and an indistinct basal stratum. Intine stratified, with tubules and fibrillar.

S. martianus (Fig. 3 c). Pollen grains 3-colporate, oblate spheroidal, 35 × 37 µm. Amb triangular. Apocolpium diameter 10 µm. – Colpi 28 × 4 µm. Colpus membrane with a median split. Ora distinct, rounded to lolongate, relatively large (6 × 4 µm). – Exine 3 µm thick. Sexine as thick as nexine, scabrate; uneven, depressed and granular at mesocolpia, with densely spaced perforations and slits (SEM).

Cerbereae. *Cerbera floribunda* (Fig. 3 e, f). Pollen grains 3-colporate, oblate spheroidal, 68 × 73 µm. Amb rounded-triangular. Apocolpium diameter 18 µm. – Colpi 50 × 2 µm, widened at centre, with smooth colpus margins (c. 5 µm wide). Ora lolongate (10 × 2 µm). – Exine 3 – 4 µm thick. Sexine as thick as nexine, scabrate; distinctly uneven with numerous perforations and slits, smooth and even at the colpus margins and the polar areas (SEM). Inside of exine with numerous granules (SEM).

Fig. 4. *a Allamanda puberula* var. *glabrata.* Part of a 3-colporate pollen grain showing a relatively short colpus with rounded ends, a granular membrane and a lolongate os in the centre. Sexine smoth, densely perforate. SEM, × 2 300. *b A. cathartica.* Exine consisting of an unevenly thick, tectal stratum, a granular infratectal stratum with larger granules towards the base, partly fused to a sole. The sole is subtended by a thin endexinous (?) stratum, and structured-fibrillar intine. TEM, × 14 000. *c Thevetia amazonica.* Fragment of a pollen showing a tectal stratum, a relatively thick, granular infratectal stratum, and a sole. Inner exine granular. SEM, × 4 900. *d T. ahouai.* Exine stratified into a discontinuous tectal stratum, a granular infratectal stratum with larger grains towards the base, a discontinuous sole and lamellar endexine. TEM, × 15 000. *e Cerbera manghas.* Exine stratified into an unevenly thick tectal stratum, a lamellar-granular infratectal stratum subtended by a thin indistinct stratum below which there is a second darkly stained (endexinous?) stratum. Intine many times thicker than exine, structured. TEM, × 22 000

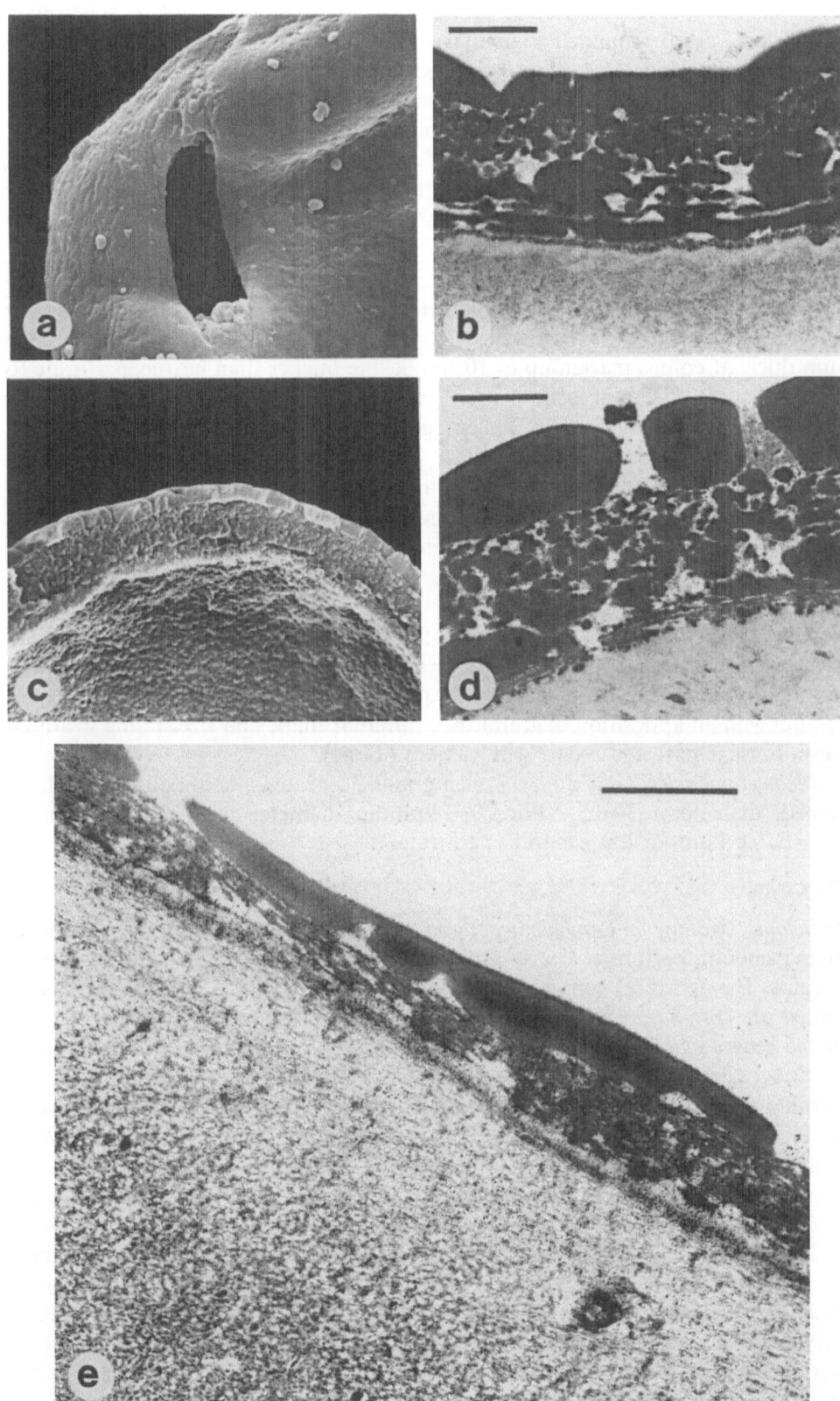

C. manghas (Fig. 4 e). Pollen grains 3-colporate, mostly oblate spheroidal, 68 × 72 µm. Amb rounded-triangular. Apocolpium diameter 20 µm. – Colpi 48 × 2 µm, with smooth, 4 – 5 µm wide margins. Ora lolongate (6 × 2 µm). – Exine 3 µm thick. Sexine as thick as nexine, scabrate perforate; stratified into a tectal stratum, a lamellar-granular infratectal stratum, subtended by a thin basal stratum. Intine structured (TEM).

Thevetia ahouai (Fig. 4 d). Exine stratified into a tectal stratum, a granular infratectal stratum with larger granules at the base, partly fused to a discontinuous sole.

T. amazonica (Fig. 4 c). Pollen grains 3-colporate, suboblate, 64 × 76 µm. Amb rounded-triangular. Apocolpium diameter 50 – 52 µm. – Colpi 18 × 2 µm, constricted at equator, with a 5 µm wide margin. Ora lolongate (5 × 2 µm). – Exine 5 µm thick, at colpus margin up to 10 µm. Sexine thicker than nexine, perforate to microreticulate. Inside of exine densely granular (SEM).

T. peruviana (Fig. 3 g, h). Pollen grains 3-colporate, oblate spheroidal, 70 × 75 µm. Amb rounded-triangular. Apocolpium diameter 38 µm. – Colpi 42 × 3 µm, constricted at equator, with a 5 µm wide margin. Ora lolongate (6 × 3 µm). – Exine 5 µm thick. Sexine thicker than nexine, perforate to microreticulate. Nexine with numerous fissures (endocracks).

Apocynoideae

Nerieae. *Alafia lucida* (Fig. 2 f; Nilsson 1986). Pollen grains 3-porate, oblate spheroidal to irregularly shaped, 30 × 33 µm. – Pores 2 – 3 µm in diameter with indistinct annulus. – Exine c. 1 µm thick, smooth, perforate; stratified into an outer tectal stratum with an apposition of granules of different shape and size. Intine stratified into a fibrillar part and a part with tubules (TEM).

Pleioceras barteri (Fig. 2 c, d). Pollen grains 3-porate, spheroidal to irregularly shaped, diameter c. 30 µm. – Pores 4 – 5 µm in diameter with indistinct annulus. – Exine 1 µm, or less, smooth, perforate.

Discussion

Ambelania, Molongum (*Ambelanieae*) and *Macoubea* (*Macoubeeae*) are all similar having smooth, perforate, clearly aspidote pollen grains with protruding aperture margins. The apertures consist of very short, relatively wide colpi with large, well limited ora (Fig. 1 e – k). The pollen grains of *Tabernaemontana* (Fig. 1 a – d) and related genera in *Tabernaemontaneae* are different. Thus, the close affinity between the above three tribes suggested by Fallen (1986), is not supported by pollen morphology. Preliminary palynological results show that the *Tabernaemontaneae* are fairly uniform compared to, e.g., the *Plumeroideae* and *Cerberoideae*. With reference to the equatorial band or the distinctly lalongate ora, the *Tabernaemontaneae* may be regarded as more primitive than the *Ambelanieae* and the *Macoubeeae* which have more specialized apertures (cf. Fallen 1986).

The *Plumerieae – Catharanthinae* are pollen-morphologically heterogeneous. *Rhazya* and *Amsonia* (Fig. 2 g) have similar pollen grains including exine ultrastructure (Nilsson 1986), while *Catharanthus* (Fig. 2 i; Nilsson 1986) differs in several respects. The genus *Vinca* appears totally different, in particular in its sporoderm ultrastructure. The statement by Plaizier (1981) of close relationship between *Catharanthus* and *Vinca* remains unsupported from pollen morphology.

The genus *Holarrhena* (*Plumerieae: Holarrheninae*) has smooth, 3-porate pollen grains (Fig. 2 a, b, e). HUANG (1986) reported pantoporate pollen grains in *H. antidysenterica*. As a rule the *Plumeroideae* and the *Cerberoideae* have 3-colporate pollen grains, while the *Apocynoideae* have 3-porate ones. The pollen grains of *Holarrhena* much resemble those of *Pleioceras* (Fig. 2 c, d) and *Alafia* (Fig. 2 f), also in their ultrastructure. The similarity between *Holarrhena* and some *Apocynoideae* as to pollen, seed morphology and aestivation (DE KRUIF 1981), speaks in favour of placing *Holarrhena* in the *Apocynoideae*.

The taxonomic position of *Allamanda* (*Allamandeae*) is doubtful. Its 3-colporate pollen grains (Fig. 4 a) suggest either *Plumerioideae* or *Cerberoideae*, not *Apocynoideae* (ALLORGE 1975). FALLEN (1986: 279) would like to relegate *Allamanda* to a tribe of its own. However, in a previous work, FALLEN (1985) has discussed a possible relationship between *Allamanda* and the *Cerberoideae*. The exine ultrastructure of *Allamanda cathartica* (Fig. 4 b) is very similar to that of *Thevetia amazonica* (Fig. 4 c) and, in particular to *T. ahouai* (Fig. 4 d). The exine ultrastructure of *Cerbera manghas* (Fig. 4 e) is less similar. Therefore, FALLEN's assumption of relationship between the *Allamandae* and the *Cerberoideae* based on a number of common features (FALLEN 1985: 578) has pollen morphological support.

It has been suggested that the *Cerberoideae* may have originated from rauvolfioid ancestors with *Kopsia* and *Ochrosia* as surviving links (FALLEN 1986: 279). With reference to pollen, *Ochrosia* (*Rauvolfieae: Ochrosiinae;* Fig. 3 a) differs both from *Cerbera* (Fig. 3 e, f) and *Thevetia* (Fig. 3 g, h); the latter genera are also distinguishable from each other by their outer and inner exine morphology. Furtheron, *Kopsia* (*Rauvolfieae, Vallesiinae;* Fig. 3 b) and *Skytanthus* (Fig. 3 d) have quite different sporoderm ultrastructures. Thus, the hypothesis of linking *Kopsia* and *Ochrosia* with the *Cerberoideae* appears unsupported.

The tribe *Tabernaemontaneae* (*Tabernaemontanoideae*) of the subfam. *Plumeroideae* is regarded as more advanced than other tribes of the *Plumeroideae* and the *Cerberoideae*. The *Apocynoideae* are regarded as most advanced (FALLEN 1986). Generally, the pollen morphology also favours the *Apocynoideae* as being comparatively advanced, particularly in respect to the type and construction of their apertures (annulate pores).

Sincere thanks are due to the Directors and Curators of the various herbaria for providing polliniferous material. For help in technical matters and in preparing the manuscript I am indebted to the staff members of the Palynological laboratory, Dr B. RAJ, Mrs E. GRAFSTRÖM, Mr M. HELLBOM, Mr N. EL-BAGHDADY and Mr J. PACULL. Mrs A. MELLIN typed the manuscript.

References

ALLORGE, L., 1975: Rattachement de la tribu des Allamandées aux Echitoidées (Apocynacées). – Adansonia, ser. 2, **15**: 273–276.

HUANG, T. C., 1986: The *Apocynaceae* of Taiwan. 2. A palynological study. – Sci. Rep. Tohoku Univ. 4th. ser (Biology) **39**: 75–102.

FALLEN, M. E., 1985: The gynoecial development and systematic position of *Allamanda* (*Apocynaceae*). – Amer. J. Bot. **72**: 572–579.

– 1986: Floral structure in the *Apocynaceae:* morphological, functional, and evolutionary aspects. – Bot. Jahrb. Syst. **106**: 245–286.

De Kruif, A. P. M., 1981: A revision of *Holarrhena* R. Br. (*Apocynaceae*). − Meded. Landbouwhogeschool Wageningen **81** (2): 1−40.

Nilsson, S., 1986: The significance of pollen morphology in the *Apocynaceae*. − In Blackmore, S., Ferguson, I. K., (Eds.): Pollen and spores: form and function. − J. Linn. Soc. Symp. Ser. **12**: 359−374.

− Van Campo, M., 1978: Pollen morphological studies in the *Apocynaceae* with special reference to fine structure. − 4th Int. Palynol. Confer. Lucknow (1976−77) **1**: 250−254.

Plaizier, A. C., 1981: A revision of *Catharanthus roseus* (L.) G. Don. (*Apocynaceae*). − Meded. Landbouwhogeschool Wageningen **81** (9): 1−12.

Spurr, A. R., 1969: A low-viscosity epoxy resin embedding medium for electron microscopy. − J. Ultrastruct. Res. **26**: 31−43.

Van Campo, M., Nilsson, S., Leeuwenberg, A. J. M., 1979: Palynotaxonomic studies in *Tabernaemontana* L. sensu lato (*Apocynaceae*). − Grana **18**: 5−14.

Address of the author: Dr Siwert Nilsson, Swedish Museum of Natural History, Palynological Laboratory, S-10405 Stockholm, Sweden.

Pl. Syst. Evol. [Suppl. 5], 103–121 (1990)

Pollen morphology of the *Cordioideae* (*Boraginaceae*): *Auxemma*, *Cordia*, and *Patagonula*

Joan W. Nowicke and J. S. Miller

Received November 10, 1987

Key words: Angiosperms, *Boraginaceae*, *Cordia*, *Auxemma*, *Patagonula*. – Pollen morphology, infrageneric classification, heterostyly.

Abstract: The subfam. *Cordioideae* comprises three genera: *Auxemma* (2 spp.), *Cordia* (c. 300 spp.), and *Patagonula* (2 spp.). Pollen of 47 species was examined in LM and SEM, and a selected group in TEM. *Auxemma* and *Patagonula* are 3-colporate and variously rugulose or irregularly striate. In *Cordia* 43 species were examined, including both forms of seven heterostyled ones. Six pollen types were found: sect. *Cordia* is 3-colpor(oid)ate with a striate-reticulate tectum; sect. *Varronia* is 3-porate reticulate; sects. *Gerascanthus*, *Myxa*, *Rhabdocalyx*, and *Superbiflorae* are 3-colpor(oid)ate and variously spinulose; *C. aurantiaca* and *C. taguahyensis* are 3-colpor(oid)ate and clavate; *C. lauta* is 3-porate and conspicuously clavate; and *C. bordasii* is 3-colporate and "rugulose". In TEM most cordias have a disrupted foot layer/endexine interface. Pollen from the long and short styled plants is only weakly dimorphic in SEM and TEM. Most *Cordioideae* can be distinguished from most *Ehretioideae* by the pseudocolpi or pseudocolpoid depressions and rugulose tecta in the latter subfamily. The problems of measuring pollen grains are also discussed.

The *Boraginaceae* are widely distributed family of about 100 genera and 2000 species circumscribed by typically sympetalous flowers, helicoid inflorescences, and trichomes with cystolith-like bases. Cronquist (1981) assigned the family to the *Lamiales* along with three others, *Lamiaceae, Lennoaceae*, and *Verbenaceae*. Cantino (1982) however, has shown that the *Boraginaceae* are somewhat anomalous in the *Lamiales* and, perhaps better placed in the *Polemoniales* where it has been assigned by other authors (Takhtajan 1969, Thorne 1968). Dahlgren (1975) assigned the *Boraginaceae* to the *Solanales*, a large aggregation including, among others, *Polemoniaceae, Hydrophyllaceae*, and *Convolvulaceae*.

Johnston (1930, 1952, 1956) recognized four subfamilies: *Cordioideae* with three genera; *Ehretioideae* with at least 12 genera; *Heliotropioideae* with three; and *Boraginoideae* with at least 40 genera. Later *Wellstedtioideae* with one species was added.

More recently, Takhtajan (1987) elevated three of the above subfamilies to family status: *Ehretiaceae* with *Bourreria, Coldenia, Cortesia, Ehretia, Halgania, Pteleocarpa, Rochefortia*, and *Rotula*; *Cordiaceae* with *Cordia, Varronia, Sebestena, Patagonula, Auxemma*, and possibly *Saccellium*; and *Wellstedtiaceae* with only

Wellstedtia. The *Boraginaceae* was then limited to *Heliotropioideae* and *Boragi-noideae.*

In this paper we view the *Cordioideae* as comprising three woody genera of very unequal size, *Auxemma* MIERS (2 species), *Cordia* L. (c. 300), and *Patagonula* L. (2). TAKHTAJAN's recent concept of *Cordiaceae/Cordioideae* is not very different from JOHNSTON's since both *Varronia* and *Sebestena* are segregated from *Cordia.*

The pollen morphology of the *Boraginaceae* is known to be highly diverse (AVETISIAN 1956, NOWICKE & RIDGWAY 1973, NOWICKE & SKVARLA 1974, CLARKE 1977, DIEZ 1984, NOWICKE & MILLER unpubl. data). Although JOHNSTON did not utilize pollen morphology in his early studies of the family, he did in some of his later ones, especially in the *Lithospermeae* (1952). In a study based primarily on light microscopy, NOWICKE & RIDGWAY (1973) found that pollen morphology correlated well with sectional classification in *Cordia*, whereas in *Tournefortia* (NOWICKE & SKVARLA 1974) diverse pollen types crossed sectional boundaries. Despite the apparent potential of pollen morphology in the classification of the *Boraginaceae*, the studies mentioned above cover few genera, and among the genera of the three tropical subfamilies only *Tournefortia* was examined in depth in SEM and TEM.

This paper is the first of a series on the pollen morphology of *Boraginaceae*, emphasizing and integrating data from TEM with that from LM and SEM. As part of the present study the pollen of more than 200 species of *Boraginaceae* has already been examined in LM and SEM and a selected group also in TEM.

Materials and methods

The species examined, with authors, all voucher data, figure citation(s), and measurements are given in Table 1.

Anthers were removed from herbarium specimens (Table 1) and acetolyzed for all preparations, LM, SEM, and TEM. Slides for LM were made with glycerin jelly and sealed with wax. Material for SEM was coated with carbon, then gold-palladium and photographed with a Hitachi S-570. Material for TEM was fixed with osmium tetroxide, stained with uranyl acetate, dehydrated, and embedded in Spurr's medium. After sectioning with diamond knives, the sections were stained with lead citrate, and photographed with a JEOL 100 CX TEM. For a more detailed account of TEM preparation, see NOWICKE & al. (1986).

Slides for LM and the SEM/TEM data are deposited at the Palynological Laboratory, Botany Department, National Museum of Natural History, Washington, DC.

Measurements given in Table 1 are equatorial diameters based on ten acetolyzed grains in LM, unless otherwise noted. In those cordias with elongate colpor(oid)ate apertures, acetolysis produces artifacts in shape and thus in size. Aperture rupture produces oblate shapes, or the grains collapse along the long axes of the colpi producing prolate grains. The measurement of some species, mostly in sects. *Cordia* and *Superbiflorae,* had to be taken from SEMs or from unacetolyzed grains in ALEXANDER's (1969) stain. However, when measurements of fully expanded, yet intact, acetolyzed grains were compared between LM and SEM, there was a size discrepancy. The grains in SEM were always smaller (see *C. gerascanthus*). The SEM preparation does subject grains to vacuums and this might cause shrinkage, but in some cases the measurements from LM did not even overlap with those from SEM (see *C. curbeloi*). A larger sample size might alleviate the differences. For all the above reasons, the measurements given in Table 1 should be treated with reserve. Grains in ALEXANDER's stain were mostly spheroidal, in SEM and LM most were oriented in polar view, and it was decided to record the equatorial dimensions.

Table 1. List of *Cordioideae* species examined with accession, vouchers, and pollen grain diameters in µm. [1] Less than ten grains; [2] measurements taken from SEM photographs; [3] measurements taken from unacetolyzed grains in ALEXANDER's stain

Species	Collection	Location	Figures	Equatorial diameter		
				low	mean	high
Auxemma MIERS (rugulose or irregularly striate tecta)						
Auxemma glazioviana TAUBERT	TAVARES 668 US	Brazil	1–3	(25)	27	(30)
A. oncocalyx (ALLEMAO) BAILLON	CUTLER 8098 US	Brazil	5–6, 63			
	GENTRY & al. 50087 MO	Brazil	4	(15)	18	(22)
Cordia L.						
Sect. *Cordia* (striate/reticulate tecta)						
Cordia boissieri A. DC.	CORRELL 14883 US (Long style)	Texas		(34)	35	(39)
	THARP 3678 US (Short style)	Texas		(39)	42	(46)
C. curbeloi ALAIN	Bro. LEON & SEIFRIZ 18101 US	Cuba		(52) (39)	60 44	(65) (47)[1, 2]
C. fitchii URBAN	EKMAN 10926 US	Dominican Republic	(58)	64	(73)	
C. rickseckeri MILLSP.	LITTLE & WOODBURY 23727 US	Virgin Islands	7–9	(44)	49	(55)[2]
C. sebestena L.	CORRELL 48366 US (Long style)	Bahama Islands		(34)	37	(42)[1, 2]
	KILLIP & A. C. SMITH 14111 US (short style)	Colombia		(48)	50	(52)[1, 2]
C. seleriana FERNALD	MILLER & al. 436 MO (Short style)	Mexico	12–14	(31)	34	(36)
	MILLER & TENORIO 510 MO (Long style)	Mexico	15	(26)	28	(31)[1]
C. subcordata LAM.	BALL 12 US (Long style)	Fanning Island	10–11	(35)	40	(44)[2]
C. truncatifolia BARTLETT	MILLER & al. 1056 MO	Nicaragua		(29)	34	(39)
Sect. *Gerascanthus* (BROWNE) G. DON. (spinulose tecta)						
Cordia alliodora CHAM.	ROSE & al. 3320 US	Antigua	28–31	(31)	33	(35)[3]
C. gerascanthus L.	JACK 8225 US	Cuba	36	(30) (26)	33 29	(35) (32)[2]
C. globulifera I. M. JOHNSTON	MILLER & TENORIO 662 MO (Long style)	Mexico		(31)	33	(34)
	MILLER & TENORIO 667 MO (Short style)	Mexico	32	(33)	36	(39)

Table 1 (continued)

Species	Collection	Location	Figures	Equatorial diameter		
				low	mean	high
Sect. *Myxa* (ENDL.) DC. (spinulose tecta)						
Cordia abyssinica R. BR.	BURGER 390 US	Ethiopia	26, 27	(49)	55	(60)[3]
C. dentata POIRET	MILLER & al. 438 MO (Long style)	Mexico	22, 23	(34)	37	(39)
	MILLER & al. 439 MO (Short style)	Mexico		(37)	39	(43)
C. dichotoma FORSTER	SHIU YING HU 13267 US	Hong Kong		aberrant		
C. lutea LAM.	E. F. ANDERSON 2418 MO	Ecuador	24, 25	(49)	53	(57)
C. myxa L.	CHASE 7898 MO	S. Rhodesia		sterile		
C. nodosa LAM.	STEINBACH 435 US	Bolivia	17, 18	(39)	42	(44)
C. oblongifolia HOCHST.	HUBER 621 US	Sri Lanka	19	(34)	36	(40)[1,2]
C. ovalis R. BR.	M. RICHARDS 21601 MO	Tanzania		insufficient material		
C. pilosissima BAKER	DRUMMOND & COOKSON 6761 MO	Rhodesia		(39)	44	(46)[1]
C. platythyrsa BAKER	LOUIS, s. n. 23-VI-1936 US	Zaire	16	insufficient material		
C. wallichii G. DON.	SALDANHA 16692 US	India	20, 21	(44)	46	(48)[1,2]
Sect. *Rhabdocalyx* A. DC. (spinulose tecta)						
Cordia elaeagnoides DC.	ANDERSON & LASKOWSKI 3774 US (Long style)	Mexico	34, 35	(25)	27	(29)[3]
	HINTON & al. 13293 US (Short style)	Mexico		(30)	31	(33)[3]
C. varronifolia I. M. JOHNSTON	HUTCHISON & WRIGHT 6774 US	Peru	33	(33)	34	(36)
Sect. *Superbiflorae* TARODA (spinulose tecta; clavate in *C. taguahyensis*)						
Cordia aberrans I. M. JOHNSTON	RESTINGA-1 1325 US	Brazil		(48)	51	(54)[1,2]
C. anabaptista CHAM.	BELEM & PINHEIRO 3293 MO	Brazil		(40)	41	(44)[1,2]
C. rufescens A. DC.	HATSCHBACH & GUIMARAES 25519 US	Brazil	37−39	(48)	51	(56)[2]
C. superba CHAM.	HARLEY & al. 15225 US	Brazil	42	(59)	61	(64)[1,2]
C. taguahyensis VELL.	W. R. ANDERSON 9130 MO	Brazil		(55)	60	(65)
	DAWSON 14996 MO	Brazil		(52)	56	(60)[1,2]
	PICKEL 3416 US	Brazil	40, 41	(46)	49	(54)[1,2]

Table 1 (continued)

Species	Collection	Location	Figures	Equatorial diameter		
				low	mean	high

Sect. *Varronia* (BROWNE) G. DON. (reticulate tecta; clavate in *C. lauta*)

Species	Collection	Location	Figures	low	mean	high
Cordia acuta PITTIER	CUATRECASAS 22389 US	Colombia	57	(43)	46	(48)
C. bullata (L.) R. & S.	MILLER & STEVENS 1062 MO (Long style)	Nicaragua	56	(33)	37	(40)[2]
C. chabrensis URBAN & EKMAN	Bro. LIOGIER 11567 US	Dominican Republic		(48)	52	(59)
C. curassavica (JACQ.) R. & S.	KILLIP 44810 US	Cuba	43 – 47	(40)	43	(45)[1]
C. globosa (JACQ.) HBK.	KILLIP & SMITH 20942 US	Colombia	53, 54	(36)	43	(48)
	NICHOLS 136 US	Nicaragua		(39)	44	(52)
C. grandiflora HBK.	HOLT & BLAKE 840 US	Venezuela		(49)	52	(55)
C. lauta I. M. JOHNSTON	GREGG 981 MO	Mexico	58	(88)	97	(107)[3]
	MILLER 3159 MO	Mexico		(107)	111	(115)[3]
C. oaxacana DC.	BITTMAN & al. 367 US	Mexico		(52)	61	(65)
C. polycephala (LAM.) I. M. JOHNSTON	ALLARD 13226 US (Long style)	Dominican Republic		(35)	39	(44)
	R. A. & E. S. HOWARD 9749 US (Short style)	Dominican Republic		(40)	43	(44)
	PURSELL & al. 9029 US	Venezuela	50 – 52	(34)	38	(43)
C. pringlei B. L. ROBINSON	MILLER & MORENO 1052 MO (Short style)	Nicaragua	55	(36)	40	(44)[2]
C. spinescens L.	MARTINEZ-CALDERON 1467 MO (Long style)	Mexico	48	(36)	39	(40)
	MILLER & TENORIO 591 MO (Short style)	Mexico	49	(34)	38	(42)
	MORENO 9558 MO (Short style)	Nicaragua		(40)	43	(46)
	NEILL 7358 MO (Long style)	Nicaragua		(36)	39	(43)

Section uncertain

Species	Collection	Location	Figures	low	mean	high
C. aurantiaca BAKER (Clavate tectum)	THOMAS 3487 MO	Cameroon		(59)	66	(71)
C. bordasii SCHINI (Rugulose tectum)	CARDENAS 6202 US	Bolivia		(27)	29	(33)
C. decandra H. & A. (Spinulose tectum)	MORONG 1194 US	Chile		(27)	30	(33)

Patagonula L. (Rugulose tecta)

Species	Collection	Location	Figures	low	mean	high
Patagonula americana L.	SMITH & REITZ 12658 US	Brazil	59 – 62	(14)	15	(16)[1, 2]
P. bahiensis MORICAND	SCHREINER 14826 US	Brazil		(16)	17	(19)[1, 2]

Results

Auxemma Miers, Figs. 1 – 6, and 63. Grains spheroidal, oblate spheroidal; 3-colpor(oid)ate, the colpi long, the margins granular; the tectum irregularly striate or rugate-lirate; (in TEM) the endexine well-defined, becoming thicker and more granular beneath the apertures; the foot layer irregular; the columellae reduced; the tectum incomplete and irregular, reflecting the morphology depicted in SEM.

Auxemma consists of two species in Brazil, *A. glazioviana* (Figs. 1 – 3) and *A. oncocalyx* (Figs. 4 – 6, and 63).

In the two collections of *A. oncocalyx* (Table 1) the pollen was uniform – a regulose tectum having distinct elements; whereas in *A. glazioviana*, the tectum was irregularly striate and quite variable but none of the variants were similar to the tectum found in *A. oncocalyx*.

Cordia L., Figs. 7 – 58. Grains oblate spheroidal, spheroidal, to prolate spheroidal; 3-porate, or 3-colpor(oid)ate, the colpi variable in length, the margins gran-

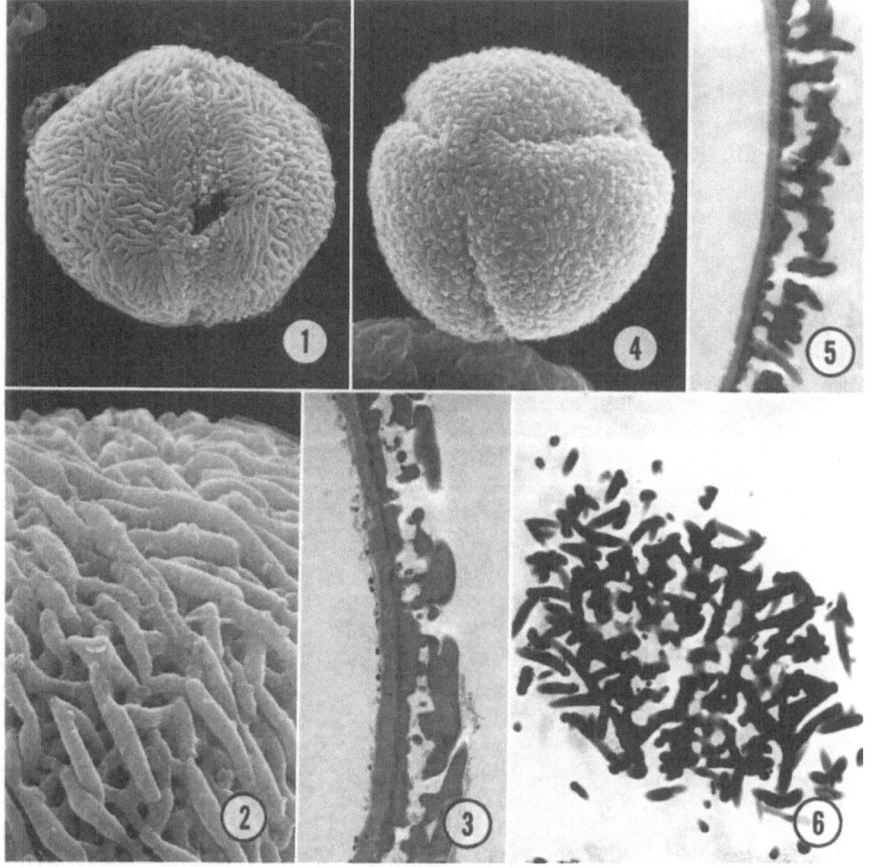

Figs. 1 – 6. SEM and TEM of *Auxemma* pollen grains. – Figs. 1 – 3. *A. oncocalyx*; Fig. 1. Equatorial view, SEM, × 1 850. Fig. 2. High mag of tectum of another grain, SEM, × 6 300. Fig. 3. Oblique radial section, irregular tectum reflects morphology depicted in Figs. 1 and 2. TEM, × 8 300. – Figs. 4 – 6. *A. glazioviana*. Fig. 4. Slightly oblique polar view, SEM, × 2 690. Fig. 5. Slightly oblique radial section, note irregular foot layer, tectum here and in Fig. 6 reflects morphology depicted in SEM Fig. 4, TEM, × 8 400. Fig. 6. Tangential section, see legend of Fig. 5, TEM, × 6 720

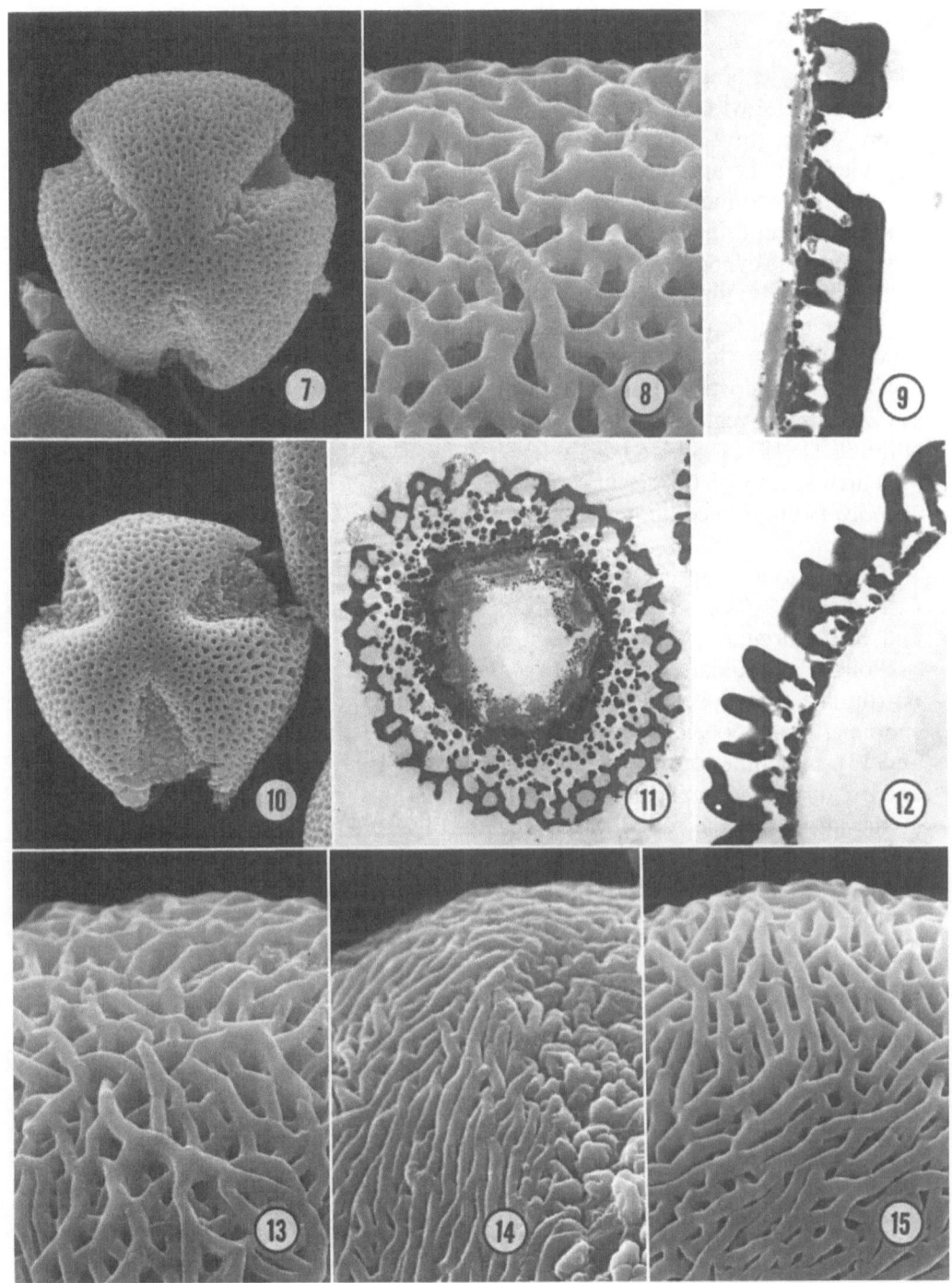

Figs. 7 – 15. SEM and TEM of pollen in *Cordia* sect. *Cordia*. – Figs. 7 – 9. *C. rickseckeri*. Fig. 7. Polar view, SEM, ×9 350. Fig. 8. Tectum, SEM, ×6 375. Fig. 9. Radial section, note irregularity of foot layer and that some columellae extend as such to the irregular endexine. TEM, ×8 500. – Figs. 10 and 11. *C. subcordata*. Fig. 10. SEM, ×1 020. Fig. 11. Tangential section, note irregular boundaries of foot layer and endexine; compare columellae with those in sect. *Varronia* (Figs. 45 and 52), TEM, ×2 800. – Figs. 12 – 15. *C. seleriana*. Figs. 12 – 14. Short style. Fig. 12. Oblique section, note irregular foot layer and apparent absence of endexine. TEM, ×7 055. Fig. 13. Tectum, compare with Fig. 14 from same collection, SEM, ×6 375. Fig. 14. Tectum, SEM, ×6 375. Fig. 15. Long style form, SEM, ×6 375

ular, usually with an irregular endoapertural opening; the tectum reticulate, variously spinulose with minute perforations, striate-reticulate, rarely clavate, or very rarely striate; (in TEM) the endexine very thin but thicker and with ectexinous granules near the apertures; the foot layer mostly irregular, either with channels, or sometimes columellae appear to extend as such to the endexine; the interface between the endexine and foot layer mostly irregular, especially near the apertures; the columellae cylindrical and well developed, or irregular and poorly defined; the tectum variable, almost complete with minute perforations and then variously spinulose or more rarely clavate or partially so, or incomplete and then striate-reticulate, reticulate, or very rarely microreticulate and clavate.

One characteristic that appears to unite much of the diverse pollen assemblage in *Cordia* is an irregular/disrupted interface between the endexine and the foot layer (Figs. 9, 11, 18, 21, 39, and 54).

Three species of *Cordia* have clavate pollen: *C. lauta* (sect. *Varronia*), *C. aurantiaca* (section uncertain); and some collections of *C. taguahyensis* (sect. *Superbiflorae*).

Sect. *Cordia*. Species examined: *C. boissieri*, l.s. (long style), s.s. (short style), *C. curbeloi, C. fitchii, C. rickseckeri, C. sebestena*, l.s., s.s., *C. seleriana*, l.s., s.s., and *C. subcordata* (Figs. 7 – 15).

Pollen spheroidal, oblate spheroidal, sometimes prolate-spheroidal; 3-colpor(oid)ate, the colpi elongate, the colpus membrane granular (Fig. 14), an irregular endoaperture may be enhanced by acetolysis; the tectum striate-reticulate (Figs. 8 and 13), sometimes mostly striate (Figs. 14 and 15); (in TEM) the endexine thin and irregular in non-aperture regions (Fig. 9) but thicker and incorporating ectexine at the apertures; the foot layer irregular (Fig. 9), sometimes channeled or with columellae extending as such to the endexine (Fig. 9); the columellae well-defined, elongate, mostly circular in tangential section (Fig. 11); the tectum thick but incomplete (Fig. 9).

The striate-reticulate tectum of sect. *Cordia* (Figs. 8 and 13 – 15) distinguishes it from all other sections: sects. *Gerascanthus, Myxa, Rhabdocalyx*, and *Superbiflorae* have spinulose tecta (Figs. 19, 20, 24, 26, 29, and 38); sect. *Varronia* has reticulate tecta (Figs. 43, 44, 48 – 50, 53, and 55 – 58).

Cordia boissieri, C. seleriana (Figs. 13 – 15), and *C. truncatifolia* have tecta that are more striate than reticulate, whereas in *C. curbeloi, C. fitchii, C. rickerseckeri* (Figs. 7 and 8), *C. sebestena* and *C. subcordata* (Fig. 10) the tecta are more reticulate than striate.

In *Cordia curbeloi* the tectum at the poles had fewer but larger perforations.

Sect. *Varronia*. Species examined: *C. acuta, C. bullata, C. chabrensis, C. curassavica, C. globosa, C. grandiflora, C. lauta, C. polycephala*, l.s., s.s., *C. pringlei, C. spinescens*, l.s., s.s. (Figs. 43 – 58).

Grains mostly oblate spheroidal, sometimes spheroidal; 3-porate, the pores small (Figs. 43, 53, 56, and 57), aperture membrane reflects tectum; the tectum reticulate with the muri/lumina of variable size (Figs. 53 and 57), the muri sparsely and uniformly spinulose (Figs. 48, 49, and 55), in one species (*C. lauta*) the tectum microreticulate and clavate (Fig. 58); (in TEM) the endexine thin, irregular and granular (Figs. 56, 51, and 54), thickening slightly near the apertures (Fig. 54); the foot layer thin (Figs. 46 and 47), frequently highly irregular at the interface with the endexine (Fig. 54); the columellae well defined, circular in tangential section

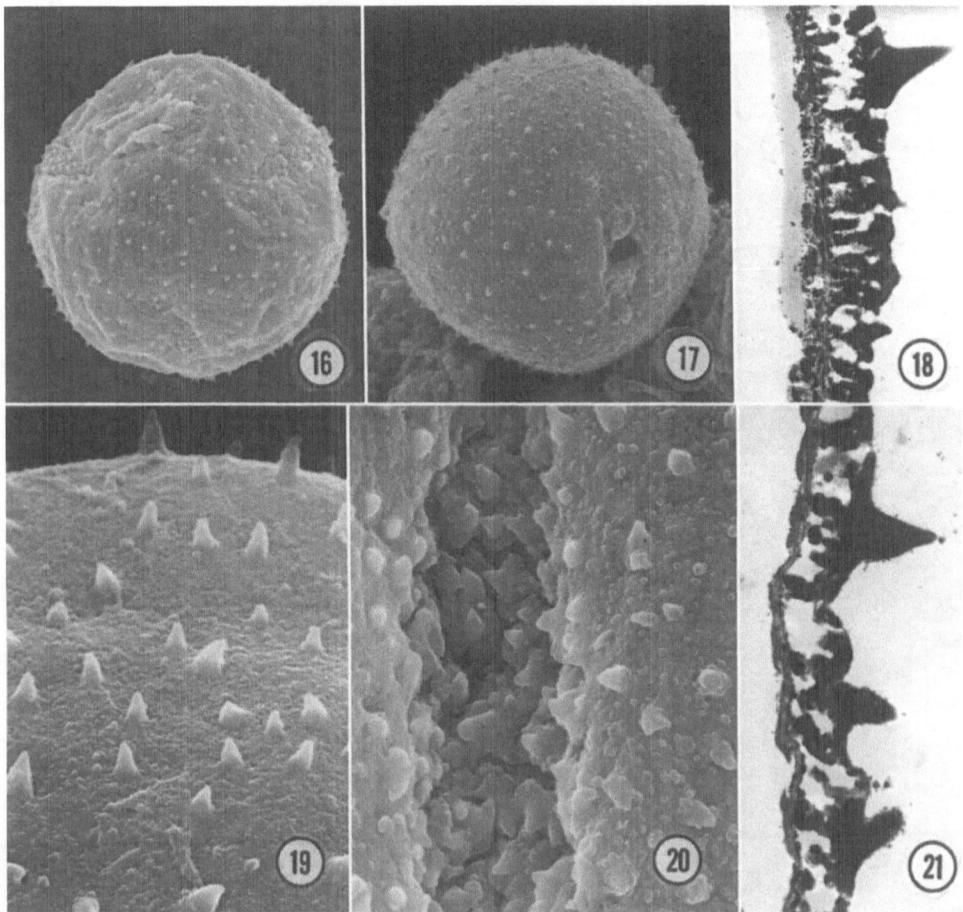

Figs. 16–21. SEM and TEM of pollen in *Cordia* sect. *Myxa*. – Fig. 16. *C. platythyrsa*, SEM, × 1090. – Figs. 17 and 18. *C. nodosa*. Fig. 17. Whole grain, note short colpus, SEM, × 1260. Fig. 18. More or less radial section, note disrupted foot layer and irregular columellae, TEM, × 6970. – Fig. 19. *C. oblongifolia*, SEM, × 6300. – Figs. 20 and 21. *C. wallichii*. Fig. 20. Tectum along colpus, SEM, × 4200. Fig. 21. More or less radial section, note disrupted foot layer/endexine boundary. TEM, × 8400

(Figs. 45 and 52); the tectum incomplete (Figs. 56, 47, 51, and 54), the muri sometimes channeled.

The pollen of sect. *Varronia* can be distinguished from the remaining sections by porate apertures and/or the reticulate exine. Simple porate apertures are rare in the *Boraginaceae*, but there is no evidence that the pore in sect. *Varronia* is the culmination of a gradual shortening of a colpus, or that it represents an endoaperture minus the colpus.

With the exception of *Cordia lauta*, pollen variation in this section is due largely to size differences in the muri and lumina. Of the 11 species examined in this section, *C. curassavica* (Fig. 43) and *C. globosa* (Fig. 53) have the thickest muri and smallest lumina. In fact the tectum of the later species could represent an advanced stage in the evolution of a complete tectum.

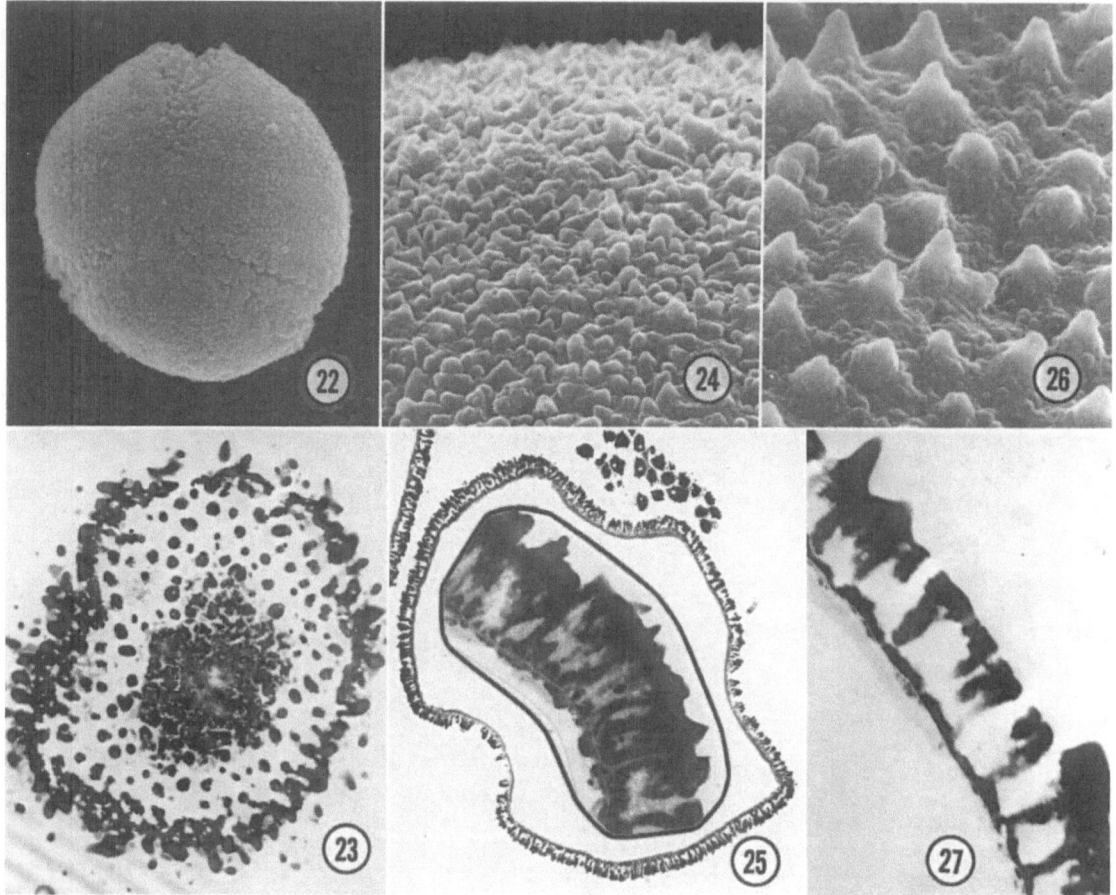

Figs. 22 – 27. SEM and TEM of pollen in *Cordia* sect. *Myxa.* – Figs. 22 and 23. *C. dentata*, long style. Fig. 22. Polar view, note elongate colpi, one of which extends to pole, SEM, × 1 445. Fig. 23. Tangential section, note well-defined columellae, TEM, × 5 610. – Figs. 24 and 25. *C. lutea*. Fig. 24. Tectum, note small, densely spaced spinules, SEM, × 6 375. Fig. 25. Oblique section of whole grain, inset, radial section showing disrupted foot layer and densely spaced columellae, TEM, × 1 360 and × 8 300. – Figs. 26 and 27. *C. abyssinica*. Fig. 26. Tectum, note surface between spinules, SEM, × 17 000. Fig. 27. Oblique radial section, note thin foot layer and perforations in tectum, TEM, × 5 610

When JOHNSTON described *Cordia lauta*, he cited the large size of the pollen with its "verrucose surface" and referred to the species as (JOHNSTON 1956: 289) "one of the most distinct members of the genus that has been discovered in America during the past quarter century." In fact *C. lauta* is the most distinctive species in *Cordia* by having the largest flowers, corollas 6 to 8 cm long, and the largest pollen, 88 – 115 µm in diameter. JOHNSTON assigned this species to sect. *Varronia*, and although the pollen is porate it still represents a discordant element with its very large size, thin exine, and microreticulate tectum with numerous clavae.

The pollen of sects. *Gerascanthus, Myxa, Rhabdocalyx,* and *Superbiflorae* is sufficiently similar to be discussed together.

Sect. *Gerascanthus. Species examined: C. alliodora, C. gerascanthus, C. globulifera,* l.s., s.s. (Figs. 28 – 32, and 36).

Sect. *Myxa.* Species examined: *C. abyssinica, C. dentata* l.s., s.s., *C. lutes, C. myxa, C. nodosa, C. oblongifolia, C. wallichii* (Figs. 16 – 27).

Sect. *Rhabdocalyx.* Species examined: *C. elaeagnoides* l.s., s.s., *C. varronifolia* (Figs. 33 – 35).

Sect. *Superbiflorae.* Species examined: *C. aberrans, C. anabaptista, C. rufescens, C. superba,* and *C. taguahyensis* (Figs. 37 – 42).

Grains oblate-spheroidal to spheroidal; 3-colpor(oid)ate, the colpi of variable length (Figs. 17 vs. 28), the endoapertural opening irregular (Figs. 17 and 32); the tectum minutely perforate and variously spinulose (Figs. 19, 20, 24, 26, 29, and 38), or very rarely clavate (Figs. 40 and 41); (in TEM) the endexine thin (Figs. 31 and 34) but much thicker beneath the apertures; the foot layer thin (Fig. 27) and very irregular (Figs. 18, 21, and 39); the interface between foot layer and endexine very irregular (Figs. 18, 25 inset, and 39); the columellae irregular and densely spaced (Figs. 18, 25, and 39), or more cylindrical and more sparsely distributed (Figs. 23, 27, and 34); the tectum thin and minutely perforate (Figs. 18, 21, 27, 31, 34, and 39).

Cordia abyssinica, C. dentata, and *C. lutea* (sect. *Myxa*), all share large corollas, ribbed calyx, and a tectum composed of small spinules uniformly distributed. This tectal variant was not found in any other taxa examined. In addition, in tangential section they have more cylindrical and sparsely distributed columellae.

Cordia gerascanthus and *C. globulifera* (sect. *Gerascanthus*) have short colpi, while in *C. nodosa* (sect. *Myxa*) the colpi are both short and poorly defined.

In his treatment of *Cordia* from Brazil, Paraguay, Uruguay, and Argentina, JOHNSTON (1930) characterized sect. *Eucordia* as having loose panicles or cymes, large corollas, and fruits enclosed to some degree by an enlarged calyx. He assigned six species: *C. aberrans* (as *C. mucronata*), *C. anabaptista, C. candida, C. rufescens, C. superba,* and *C. taguahyensis.* Later, TARODA & GIBBS (1986) made *C. superba* the type for a new section, sect. *Superbiflorae,* and characterized it as consisting of the six closely related species assigned by JOHNSTON to sect. *Eucordia.* The pollen of these species does not have the striate-reticulate exine characteristic of *C. sebestena* and allied species (sect. *Cordia*), and we agree with their exclusion from sect. *Cordia.*

One collection of *C. taguahyensis* (PICKEL 3416) had a conspicuously clavate tectum and short colpi (Figs. 40 and 41). Two more collections (ANDERSON 9130, DAWSON 14996) had pollen that was more spinulose than clavate and with elongate colpi. There are size differences among the three collections, and the possibility exists that the clavate and spinulose morphologies may be linked to heterostyly. However, the flowers in these collections were either in bud or insufficient in number to determine style length with confidence. Occasional clavae were observed in *C. aberrans* and *C. rufescens,* also in sect. *Superbiflorae.*

Three species, *Cordia aurantiaca, C. bordasii,* and *C. decandra* are not assigned to section (section uncertain in Table 1). They are discussed only briefly, since they will be the subject of a separate paper.

While BAKER (1894) did not assign *Cordia aurantiaca* to a section when he described it, sect. *Myxa* would be the logical choice, since inflorescence/floral morphology eliminates sects. *Cordia* and *Varronia.* However, *C. aurantiaca* has clavate pollen in contrast to the spinulose type in sect. *Myxa.*

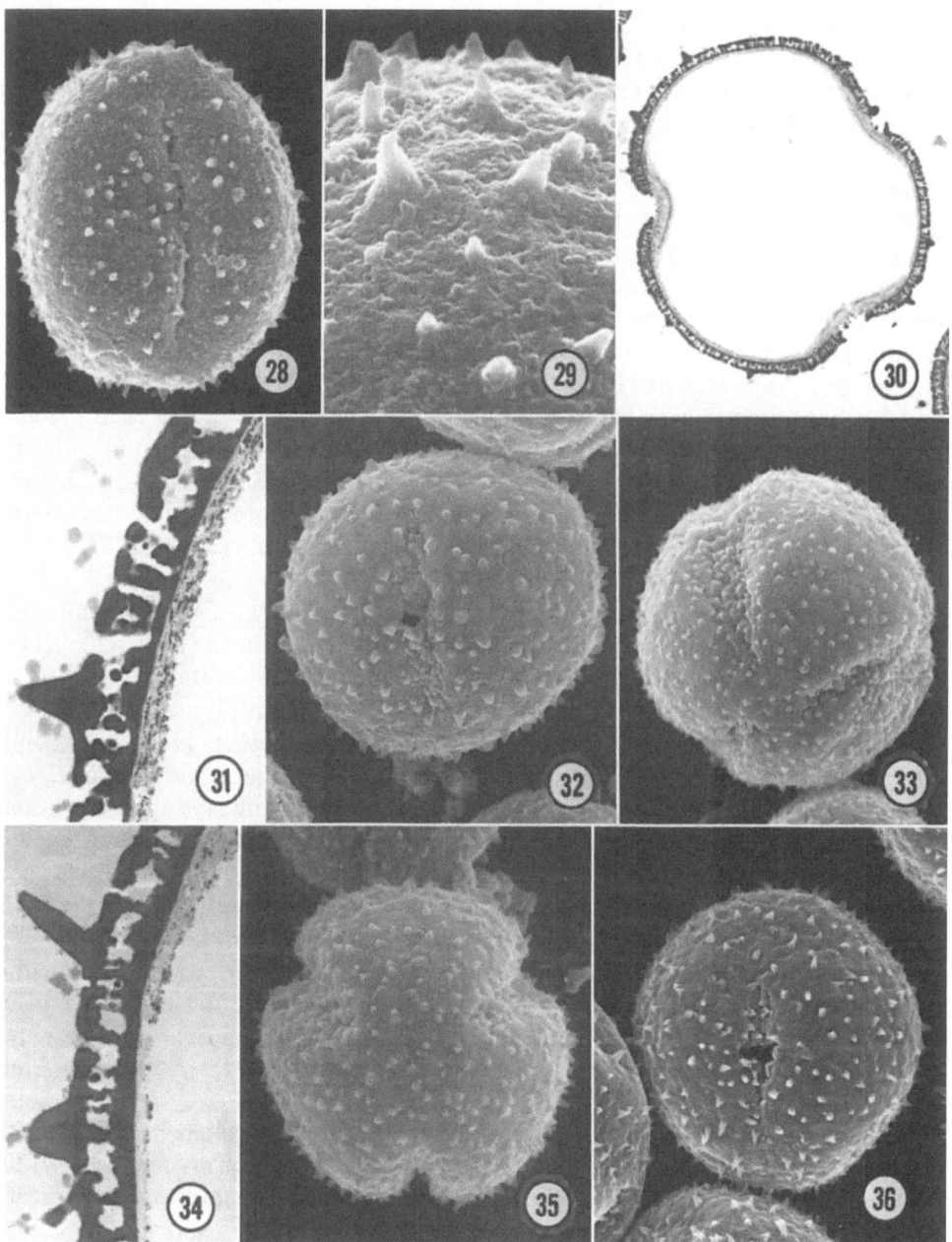

Figs. 28 – 36. SEM and TEM of *Cordia* sectt. *Gerascanthus* (Figs. 28 – 32, and 36) and *Rhabdocalyx* (Figs. 33 – 35). – Figs. 28 – 31. *C. alliodora*. Fig. 28. Whole grain, SEM, × 1 510. Fig. 29. Tectum, SEM, × 6 300. Fig. 30. Radial section whole grain, TEM, × 1 680. Fig. 31. Radial section, *C. alliodora* is one of the few species in *Cordia* where the boundary of endexine/foot layer is well defined, TEM, × 8 400. – Fig. 32. *C. globulifera*, short style, SEM, × 1 430. – Fig. 33. *C. varronifolia*, SEM, × 1 275. – Figs. 34 and 35. *C. elaeagnoides*, longstyle. Fig. 34. Radial section, TEM, × 8 400. Fig. 35. Whole grain, SEM, × 1 850. – Fig. 36. *C. gerascanthus*, note short colpus and narrow spines, SEM, × 1 430

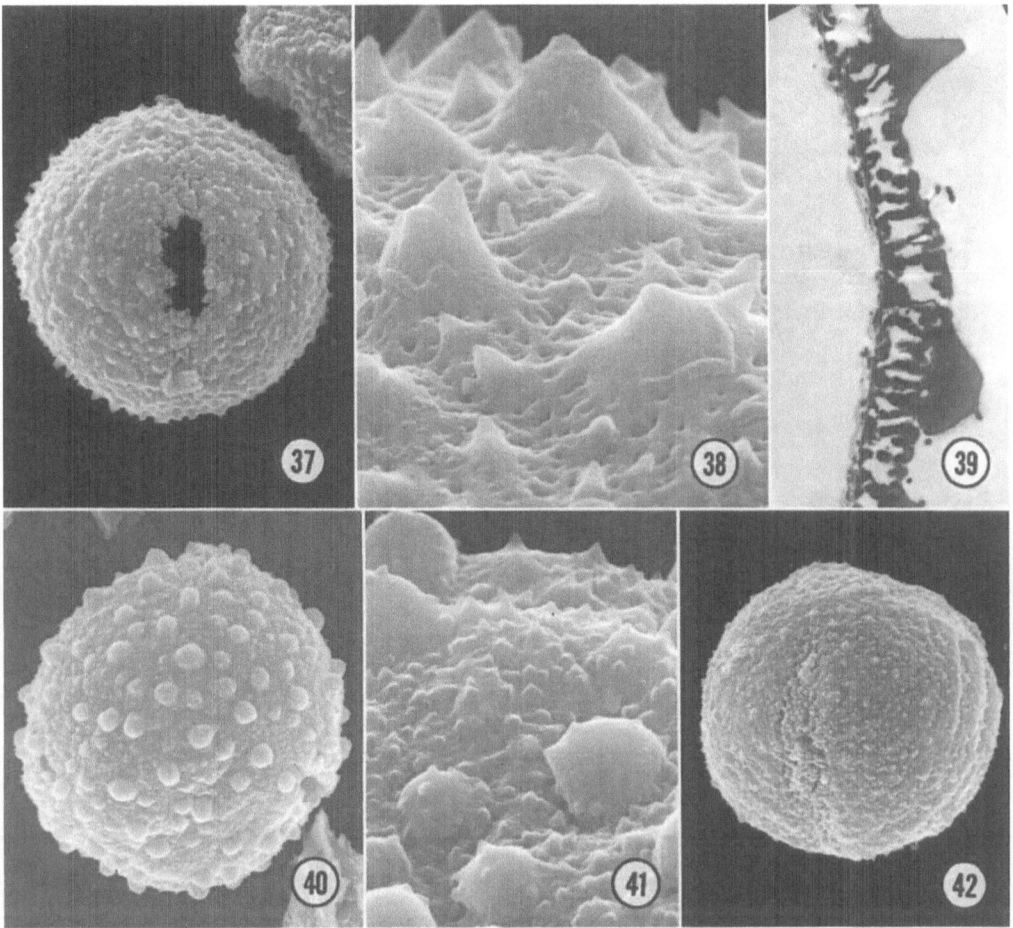

Figs. 37 – 42. SEM and TEM of pollen in *Cordia* sect. *Superbiflorae*. – Figs. 37 – 39. *C. rufescens*. Fig. 37. Whole grain, note large endoaperture and poorly defined colpus, SEM, × 850. Fig. 38. Tectum, note surface between spinules and compare with that in Fig. 41 of *C. taguahyensis* and Fig. 27 of *C. abyssinica*, SEM, × 8 500. Fig. 39. Radial section, note thin tectum, and small columellae in relation to stout spines, as well as irregular foot layer, TEM, × 5 610. – Figs. 40 and 41. *C. taguahyensis* (PICKEL 3416). Fig. 40. Polar view of grain with very short colpi, SEM, × 850. Fig. 41. Tectum showing small spinules between and upon the clavae. In other collections of *C. taguahyensis* (see Table 1) the tectum was more spinulose than clavate. SEM, × 850. – Fig. 42. *C. superba*, note poorly defined colpus, SEM, × 680

When SCHININI (1981) described *Cordia bordasii* from Paraguay, he allied it with *C. superba* in sect. *Cordia* (JOHNSTON 1930). Later, TARODA & GIBBS (1986) segregated the six Brazilian species of JOHNSTON's sect. *Cordia* as sect. *Superbiflorae*, but did not treat *C. bordasii*. *Cordia bordasii* shares a large, funnelform corolla with both sects. *Cordia* and *Superbiflorae*, as well as a conspicuously accrescent calyx more characteristic of the former section than the latter.

Cordia decandra, an unusual species of the Atacama desert in coastal Chile, was

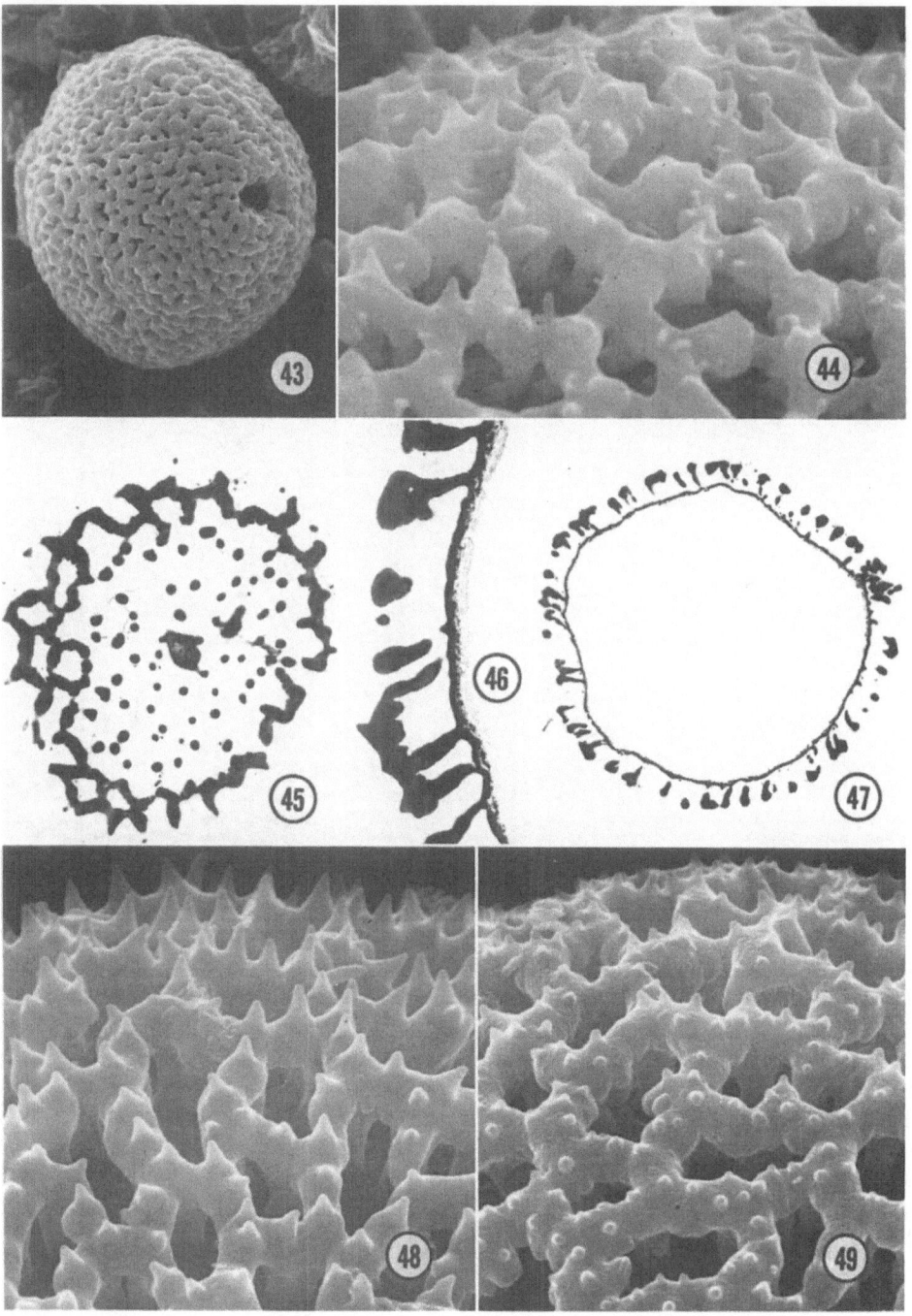

Figs. 43 – 49. SEM and TEM of pollen in *Cordia* sect. *Varronia*. – Figs. 43 – 47. *C. curassavica*. Fig. 43. Oblique view of 3-porate grain, SEM, × 1 150. Fig. 44. Tectum, SEM, × 6 325. Fig. 45. Tangential section, note sparsely distributed columellae, TEM, × 2 800. Fig. 46. Radial section, the gaps in the tectum reflect a morphology like the grain in Fig. 44 as opposed to Fig. 43, TEM, × 4 250. Fig. 47. Section of whole grain, TEM, × 1 360. – Figs. 48 and 49. *C. spinescens*. Fig. 48. Short style form, SEM, × 6 375. Fig. 49. Long style form, SEM, × 6 375

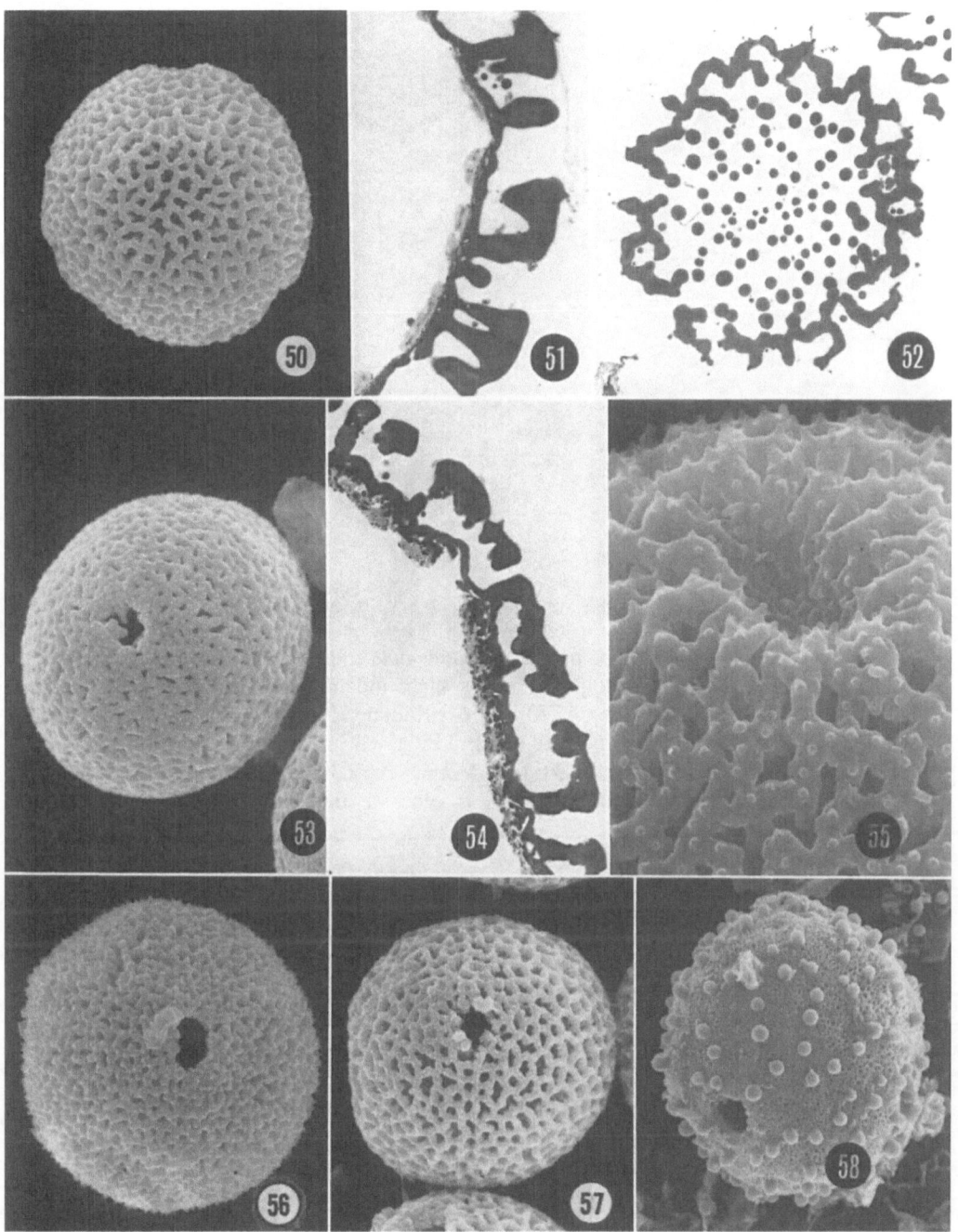

Figs. 50 – 58. SEM and TEM of pollen in *Cordia* sect. *Varronia.* – Figs. 50 – 52. *C. poly-cephala.* Fig. 50. Polar view of 3-porate grain, SEM, × 1 100. Fig. 51. Radial section, note irregular endexine, TEM, × 4 250. Fig. 52. Tangential section, note that columellae are almost perfectly round in cross section, TEM, × 2 805. – Figs. 53 and 54. *C. globosa.* Fig. 53. Note very reduced/small lumina, SEM, × 1 275. Fig. 54. Slightly oblique radial section, note disrupted foot layer/endexine, TEM, × 5 610. – Fig. 55. *C. pringlei,* aperture, SEM, × 6 375. – Fig. 56. *C. bullata,* SEM, × 1 275. – Fig. 57. *C. acuta,* SEM, × 1 105. – Fig. 58. *C. lauta,* this species has the most distinctive pollen in the genus; the tectum is finely perforate between the clavae, SEM, × 470

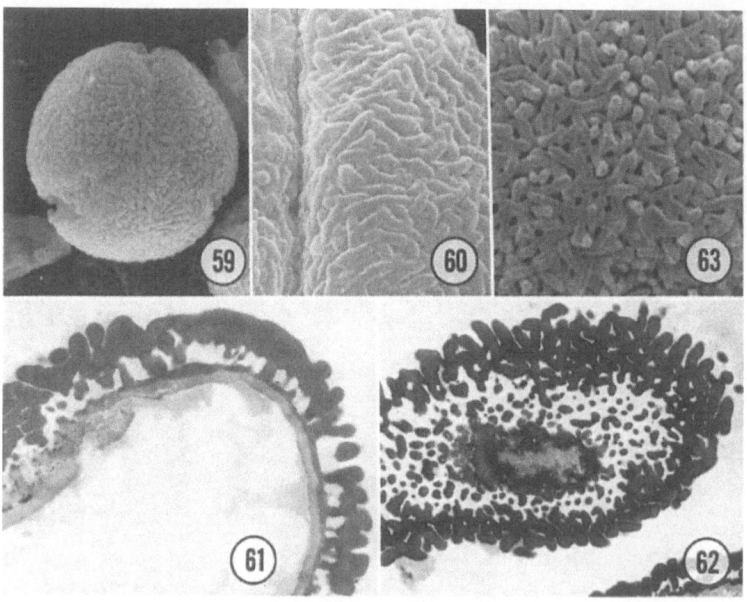

Figs. 59–63. SEM and TEM of pollen in *Patagonula* and *Auxemma*. – Figs. 59–62. *P. americana*. Fig. 59. Note that colpi extend as slight indentations almost to poles, SEM, × 1 910. Fig. 60. Tectum, SEM, × 4 770. Fig. 61. Radial section through mesocolpus, note prominence of endexine near aperture (left), TEM, × 6 360. Fig. 62. Tangential section, compare tectum with that in Fig. 61, TEM, × 5 280. – Fig. 63. *A. oncocalyx*, compare tectum with that in Fig. 60, the only real difference is a loss of some distinction of the elements, SEM, × 4 770

treated by De Candolle (1845) with taxa now placed in sect. *Cordia*. The spinulose pollen however, indicates that *C. decandra* is not closely related to sect. *Cordia*.

Heterostyly in *Cordia*. With the exception of the *Rubiaceae*, the *Boraginaceae* probably have more heterostyled species than any other angiosperm family (Ganders 1979). This is due in no small part to the high incidence of heterostyly in *Cordia*, where it is found in all sections, and in the majority of species (Darwin 1877, Opler & al. 1975, Gibbs & Taroda 1983, Miller 1985). In this study pollen of the long and short styled forms of *C. boissieri*, *C. curassavica*, *C. dentata*, *C. elaeagnoides*, *C. globulifera*, *C. polycephala*, *C. sebestena*, *C. seleriana*, and *C. spinescens* was examined in SEM. The first four species and *C. sebestena* were also examined in TEM.

To date, pollen of heterostyled cordias appears only weakly dimorphic in SEM/ TEM (Nowicke & al. 1988). In heterostyly in general, pollen from the short styled morph is larger than that of the long styled, and this appears to be the case in *Cordia*. In sect. *Cordia* pollen from the short styled morph is usually larger than the long, but there is a substantial range of variation within a sample. In *C. seleriana* there was more tectal variation within the short styled collection (Figs. 13 and 14) than between the short and long styled (Fig. 15).

In the six collections of two heterostyled species examined in sect. *Varronia*, *C. polycephala* and *C. spinescens* (Figs. 48 and 49), the long styled pollen has slightly longer spinules on the muri than does the short styled pollen.

Patagonula L., Figs. 59 – 62. Grains mostly prolate; 3-colpor(oid)ate, the colpi elongate, the endoaperture irregular in outline; the tectum rugulose; in TEM the endexine a thin layer in non-aperture regions but much thicker beneath the apertures; the foot layer thin and uniform; the columellae short and somewhat irregular; the tectum thick, incomplete, reflecting the morphology depicted in SEM.

The type of tectum found in *Patagonula americana* (Fig. 60) and *P. bahiensis* could be easily derived from that of *Auxemma glazioviana* (Fig. 4) by a partial loss of distinction of the individual elements. This fundamental tectum is of widespread occurrence in the dicotyledons, e.g., *Campanulaceae* (NOWICKE unpubl. data), *Fagaceae* (PRAGLOWSKI 1982), *Loranthaceae* (FEUER & KUIJT 1985), *Saxifragaceae* (HIDEUX & FERGUSON 1976), to name a few.

Discussion and conclusions

For the most part the pollen data suggest that the subfam. *Cordioideae* is not a closely related group of genera, in contrast to the *Ehretioideae* and the *Heliotropioideae* (NOWICKE & MILLER, unpubl. data).

Within the species examined in the *Cordioideae*, there are at least seven distinct pollen types, listed here by the taxa in which they occur: *Auxemma oncocalyx/ Patagonula*, grains 3-colpor(oid)ate and rugulose; *Auxemma glazioviana*, grains 3-colpor(oid)ate and irregularly striate. Six pollen types, two of which are restricted to a single species, are present in *Cordia*: sect. *Cordia*, 3-colpor(oid)ate and striate/ reticulate; sect. *Varronia*, 3-porate and reticulate; sects. *Myxa, Gerascanthus, Rhabdocalyx,* and *Superbiflorae*, 3-colpor(oid)ate and variously spinulose; *C. aurantiaca* and *C. taguahyensis*, 3-colpor(oid)ate and clavate; *C. lauta*, 3-porate with a microreticulate tectum having numerous clavae; *C. bordasii*, 3-colporate and rugulose tectum consisting of distinct elements.

The distinction of the two pollen types found in a genus as small as *Auxemma* is unusual. Nevertheless, *Auxemma* is a highly distinctive genus in which the calyx becomes conspicuously winged, enlarged and encloses the fruit.

Patagonula is another genus whose calyx is modified by enlarging in fruit, in this case the individual lobes become very elongate, producing a *Dipterocarpaceae*-like fruit. The pollen of the two species is very similar.

The small S. American genus *Saccelium* H. & B. which TAKHTAJAN (1987) tentatively assigned to his *Cordiaceae* has pollen very similar to *Patagonula*. Moreover, the calyx of *Saccellium* also enlarges and encloses the fruit much like the genus *Physalis* in the *Solanaceae*.

Cordia sect. *Varronia* has the most distinctive pollen due more to the porate apertures than to the reticulate exine. From our investigations and those of other workers, sect. *Varronia* and five species of *Tournefortia* (NOWICKE & SKVARLA 1974) appear to be the only taxa in *Boraginaceae* with a simple porate aperture type. Furthermore there is no evidence that these pores were derived by a progressive shortening of a colporate aperture to the area transcribed by the endoaperture. The cordias with a spinulose or striate-reticulate exine do not have well-defined endoapertures.

Although the *Cordioideae* constitute a primitive subfamily, as does the *Ehretioideae*, pollen morphology indicates that neither group is clearly ancestral to the rest of the family. Although the *Heliotropioideae* is considered to be transitional between these two subfamilies and the highly derived *Boraginoideae*, its pollen morphology is more similar to the *Ehretioideae*. In fact, the pollen morphology of the *Boraginoideae* (AVETISIAN 1956, CLARKE 1977, DIEZ 1984, NOWICKE & MILLER, unpubl. data), while strikingly diverse yet not retaining any of the types in the primitive subfamilies, would argue for a *Boraginaceae* consisting only of *Boraginoideae*.

We gratefully acknowledge the technical assistance of JANICE L. BITTNER.

References

ALEXANDER, M. P., 1969: Differential staining of aborted and nonaborted pollen. − Stain Technol. **44**: 117−122.

AVETISIAN, E. M., 1956: Morphologie des microspores de *Boraginaceae*. − Tr. Botan. Inst. Akad. Nauk Arm. S.S.R. **10**: 7−66 (in Russian).

BAKER, J. G., 1894: Diagnoses Africane I. *Boragineae*. − Kew Bull. 1894: 26−30.

CANTINO, P. D., 1982: Affinities of the *Lamiales:* a cladistic analysis. − Syst. Bot. **7**: 237−248.

CLARKE, G. C. S., 1977: *Boraginaceae*. The Northwest European pollen flora 10. − Rev. Paleobot. Palynol. **24**: 59−101.

CRONQUIST, A., 1981: An integrated system of classification of flowering plants. − New York: Columbia University Press.

DAHLGREN, R., 1975: A system of classification of the angiosperms to be used to demonstrate the distribution of characters. − Bot. Notiser **128**: 119−147.

DARWIN, C., 1877: The different forms of flowers on plants of the same species. − London: John Murray.

DE CANDOLLE, A. P., 1845: Prodromus systematis naturalis regni vegetabilis **9**: 466−501 (*Boragineae*) − Paris: Treutel et Würtz.

DIEZ, M. J., 1984: Contribucion al atlas palinologico de andalucia occidental, I. *Boraginaceae*. − Lagascalia **13**: 147−171.

FEUER, S. M., KUIJT, J., 1985: Fine structure of mistletoe pollen 6. Small-flowered neotropical *Loranthaceae*. − Ann. Missouri Bot. Gard. **72**: 187−212.

GANDERS, F. R., 1979: The biology of heterostyly. − New Zealand J. Bot. **17**: 607−635.

GIBBS, P. E., TARODA, N., 1983: Heterostyly in the *Cordia alliodora − C. trichotoma* complex in Brazil. − Rev. Braz. Bot. **6**: 1−10.

HIDEUX, M., FERGUSON, I. K., 1976: The stereostructure of the exine and its evolutionary significance of *Saxifragaceae* sensu lato. − In FERGUSON, I. K., MULLER, J. (Eds.): The evolutionary significances of the exine. − Linn. Soc. Symp. Ser. **1**: 327−377.

JOHNSTON, I. M., 1930: Studies in the *Boraginaceae* 8. Observations on the species of *Cordia* and *Tournefortia* known from Brazil, Paraguay, Uruguay and Argentina. − Contr. Gray Herb. **92**: 3−89.

− 1952: Studies in the *Boraginaceae* 23. A survey of the genus *Lithospermum*. − J. Arnold Arboretum **33**: 299−363.

− 1956: Studies in the *Boraginaceae* 28. New or otherwise interesting species from America and Asia. − J. Arnold Arbor. **37**: 288−306.

MILLER, J. S., 1985: Systematics of the genus *Cordia* (*Boraginaceae*) in Mexico and Central America. − Doctoral diss., St. Louis, MO.

NOWICKE, J. W., RIDGWAY, J. E., 1973: Pollen studies in the genus *Cordia* (*Boraginaceae*). − Am. J. Bot. **60**: 584−591.

− SKVARLA, J. J., 1974: A palynological investigation of the genus *Tournefortia* (*Boraginaceae*). − Am. J. Bot. **61**: 1021−1036.

- BITTNER, J. L., SKVARLA, J. J., 1986: *Paeonia*, exine substructure and plasma ashing. – In BLACKMORE, S., FERGUSON, I. K., (Eds.): Pollen and spores: form and function. – Linn. Soc. Symp. Ser. **12**: 81 – 95.
- MILLER, J. S., BITTNER, J. L., 1988: Pollen morphology of *Cordia sebestena* and *C. subcordata* (*Boraginaceae*). – Indian J. Palyn. **23 – 24**: 59 – 64.
- OPLER, P. A., BAKER, H. G., FRANKIE, G. W., 1975: Reproductive biology of some Costa Rican *Cordia* species (*Boraginaceae*). – Biotropica **7**: 234 – 247.
- PRAGLOWSKI, J., 1982: *Angiospermae. Fagaceae* L. *Fagoideae*. – World Pollen & Spore Flora **11**: 1 – 28.
- SCHININI, A., 1981: Contribucion a la flora del Paraguay. – Bonplandia **5** (13): 101 – 108.
- TAKHTAJAN, A., 1969: Flowering plants – origin and dispersal. – Washington: Smithsonian Institution Press.
- – 1987: Systema magnoliophytorum. – Leningrad: Nauka.
- TARODA, N., GIBBS, P. E., 1986: Studies on the genus *Cordia* L. (*Boraginaceae*) in Brazil. 1. A new infrageneric classification and conspectus. – Revta. Brasil. Bot. **9**: 31 – 42.
- THORNE, R. F., 1968: Synopsis of a putatively phylogenetic classification of the flowering plants. – Aliso **6**: 57 – 66.

Addresses of authors: JOAN W. NOWICKE, Department of Botany, Smithsonian Institution, Washington, DC 20560, U.S.A. – JAMES S. MILLER, Missouri Botanical Garden, St. Louis, MO, U.S.A.

Subject index

T. J. Mabry · G. Wagenitz (eds.)
Research Advances in the
Compositae

1990. 20 figures. V, 124 pages.
Cloth DM 138,–, öS 980,–
Reduced price for subscribers to "Plant Systematics and Evolution":
Cloth DM 124,20, öS 882,–
ISBN 3-211-82174-0

The volume presents modern research approaches for understanding evolution among members of the family *Compositae*. The symposium from which the work is derived foregrounded chemical and serological techniques, chloroplast DNA restrictive site analyses as well as classical methods for investigating the systematics of the family.

The usefulness for systematics of serological studies of the main seed storage protein of members of the *Compositae* and the way patterns of secondary plant constituents in the tribes *Vernonieae* and *Heliantheae* can aid in understanding phylogenetic relationships are emphasized.

The systematics of the large genus *Vernonia* was cladistically investigated while experimental taxonomy methods indicated that reproductive isolation is the dominant isolating factor in *Heliantheae*, with geographical isolation occurring in about two thirds of the taxa.

This valuable research volume will encourage continued research in the domain of evolutionary botany by plant systematists and phytochemists.

Plant Systematics and Evolution · Supplementum 4

Springer-Verlag Wien New York

Springer-Verlag, Mölkerbastei 5, P.O. Box 367, A-1011 Wien · Heidelberger Platz 3, D-1000 Berlin 33· 175 Fifth Avenue, New York, NY 10010, U.S.A. · 37-3, Hongo 3-chome, Bunkyo-ku, Tokyo 113, Japan.

F. Ehrendorfer (ed.)
Woody Plants —Evolution and Distribution Since the Tertiary

Proceedings of a Symposium Organized by
Deutsche Akademie der Naturforscher LEOPOLDINA in Halle/Saale,
German Democratic Republic, October 9–11, 1986

Special Edition of "Plant Systematics and Evolution", Vol. 162, 1989

1989. 146 figures. V, 329 pages. Cloth DM 310,–, öS 2170,–. ISBN 3-211-82124-4

Paleobotany has enormously expanded the documentation of fossil plant groups, floras and vegetation types, supporting its conclusions by technically much improved analyses of microfossils (pollen) and anatomical details. An increasing quantity and quality of all these informations from the geosciences is available when we follow the history of the biosphere up to the present. Simultaneously, research from the biosciences on the morphology, ecology, distribution, systematics and evolution of extant vascular plants, and on the ecogeographical differentiation of the vegetation cover of our planet, has made enormous progress. Thus, a synthetic geo-and bioscientific approach becomes more and more feasible and urgent for further advances in the many problems of common concern.

A symposium organized by the Deutsche Akademie der Naturforscher LEOPOLDINA, attractive to paleo-and neobotanists, stimulated the discussion between specialists of the two disciplines.

The main results of the symposium are now presented in this volume: Sixteen international contributions outline the current knowledge about the historical differentiation and evolution of woody plant groups and forests, covering the whole biosphere. This survey, from the beginning of the Tertiary up to the present, is a first synthesis of relevant data from the geo-and biosciences.

Springer-Verlag Wien New York

Springer-Verlag, Mölkerbastei 5, P.O. Box 367, A-1011 Wien · Heidelberger Platz 3, D-1000 Berlin 33· 175 Fifth Avenue, New York, NY 10010, U.S.A.· 37-3, Hongo 3-chome, Bunkyo-ku, Tokyo 113, Japan.

Fortschritte der Chemie organischer Naturstoffe

Progress in the Chemistry of Organic Natural Products

Founded by L. Zechmeister
Edited by W. Herz, H. Grisebach, G. W. Kirby, Ch. Tamm

The volumes of this classic series, now referred to simply as "Zechmeister" after its founder, L. Zechmeister, have appeared under the Springer Imprint ever since the series' inauguration in 1938. The volumes contain contributions on various topics related to the origin, distribution, chemistry, synthesis, biochemistry, function or use of various classes of naturally occurring substances ranging from small molecules to biopolymers.

Each contribution is written by a recognized authority in his field and provides a comprehensive and up-to-date review of the topic in question. Addressed to biologists, technologists and chemists alike, the series can be used by the expert as a source of information and literature citations and by the non-expert as a means of orientation in a rapidly developing discipline.

Springer-Verlag Wien New York

Volume 55:

1989. 41 figures. X, 208 pages.
Cloth DM 190,–, öS 1330,–
ISBN 3-211-82087-6

Contents: M. T. Davies-Coleman and D. E. A. Rivett: Naturally Occurring 6-substituted 5,6-dihydro-α-pyrones – K. Krohn: Building Blocks for the Total Synthesis of Anthracyclinones – M. Lounasmaa and J. Galambos: Indole Alkaloid Production in Catharanthus *roseus* Cell Suspension Cultures – Catherine E. James, Leslie Hough, Riaz Khan: Sucrose and Its Derivatives.

Volume 54:

1988. VII, 353 pages.
Cloth DM 320,–, öS 2240,–
ISBN 3-211-82086-8

Contents: T. Murakami and N. Tanaka: Occurrence, Structure and Taxonomic Implications of Fern Constituents.

Volume 53:

1988. 72 figures. VIII, 311 pages.
Cloth DM 275,–, öS 1930,–
ISBN 3-211-82074-4

Contents: L. F. Alves: Chemical Ecology and the Social Behavior of Animals – T. Nomura: Phenolic Compounds of the Mulberry Tree and Related Plants – A. Chimiak and M. J. Milewska: N-Hydroxyamino Acids and Their Derivatives.

Volume 52:

1987. 65 figures. VIII, 224 pages.
Cloth DM 210,–, öS 1470,–
ISBN 3-211-81989-4

Contents: U. Weiss, L. Merlini, and G. Nasini: Naturally Occurring Perylenequinones – H. Achenbach: The Pigments of the Flexirubin-Type. A Novel Class of Natural Products – T. Goto: Structure, Stability and Color Variation of Natural Anthocyanins – P. Bhattacharyya and D. P. Chakraborty: Carbazole Alkaloids.

Springer-Verlag, Mölkerbastei 5, P.O. Box 367, A-1011 Wien · Heidelberger Platz 3, D-1000 Berlin 33. 175 Fifth Avenue, New York, NY 10010, U.S.A. · 37-3, Hongo 3-chome, Bunkyo-ku, Tokyo 113, Japan.